Synthesis Lectures on Ocean Systems Engineering

Series Editor

Nikolas Xiros, University of New Orleans, New Orleans, LA, USA

The series publishes short books on state-of-the-art research and applications in related and interdependent areas of design, construction, maintenance and operation of marine vessels and structures as well as ocean and oceanic engineering.

Alexander Arnfinn Olsen

Accidental Load Analysis and Design for Offshore Structures

Alexander Arnfinn Olsen ⓘ
Southampton, UK

ISSN 2692-4420 ISSN 2692-4471 (electronic)
Synthesis Lectures on Ocean Systems Engineering
ISBN 978-3-031-74772-4 ISBN 978-3-031-74773-1 (eBook)
https://doi.org/10.1007/978-3-031-74773-1

© The Editor(s) (if applicable) and The Author(s), under exclusive license to Springer Nature Switzerland AG 2025, corrected publication 2025

This work is subject to copyright. All rights are solely and exclusively licensed by the Publisher, whether the whole or part of the material is concerned, specifically the rights of translation, reprinting, reuse of illustrations, recitation, broadcasting, reproduction on microfilms or in any other physical way, and transmission or information storage and retrieval, electronic adaptation, computer software, or by similar or dissimilar methodology now known or hereafter developed.

The use of general descriptive names, registered names, trademarks, service marks, etc. in this publication does not imply, even in the absence of a specific statement, that such names are exempt from the relevant protective laws and regulations and therefore free for general use.

The publisher, the authors and the editors are safe to assume that the advice and information in this book are believed to be true and accurate at the date of publication. Neither the publisher nor the authors or the editors give a warranty, expressed or implied, with respect to the material contained herein or for any errors or omissions that may have been made. The publisher remains neutral with regard to jurisdictional claims in published maps and institutional affiliations.

This Springer imprint is published by the registered company Springer Nature Switzerland AG
The registered company address is: Gewerbestrasse 11, 6330 Cham, Switzerland

If disposing of this product, please recycle the paper.

Preface

This short book has been written to address the process of identifying, and assessing the effects of, structural loads arising from accidental events. An essential element in the determination of accidental loads in this text is the use of risk-based assessment techniques. The traditional approach to accident-induced structural loads is the use of prescriptive criteria that are said to be primarily derived from experience and studies done for similar situations. Prescriptive criteria may be stated in terms of loading scenarios that may give specific load or pressure magnitudes, directions, durations, area of pressure application, etc. Alternatively, some prescriptive criteria may be stated in terms of presumed accident consequences, such as the loss of a major bracing member, extents of accidental collision penetration and flooding, missing mooring components, etc.

Over at least the last decade, there has been greater recognition of the use of risk-based procedures to replace or support the prescriptive criteria. For example, many classification societies have sought to develop and clarify the classification criteria for Mobile Offshore Drilling Units and Floating Production Installations. These have typically taken the form of risk-based evaluations to establish alternatives to the usual criteria given in the Class Rules and associated guidance publications. For new or novel situations, and structural types, the accidental load criteria may state primary reliance on the risk-based accidental load determination.

With the use of risk-based criteria, the guidance provided herein is meant to provide an overview of an approach that can be used to identify and assess the effects of accidental structural loads arising from any one of four hazards: dropped objects, vessel collision, fire and blast). The given methodologies should be adapted to accidental structural loads arising from other hazards as specified in the classification criteria of a particular type of offshore installation or MODU.

> Readers are cautioned that regulatory bodies having jurisdiction over the offshore installation or unit may require the use of prescriptive accidental loads criteria. Accordingly, it is the responsibility of the asset owner to discuss with the applicable authorities the acceptance of alternatives based on risk evaluations.

Southampton, UK   Alexander Arnfinn Olsen

The original version of the book has been revised. A correction to this book can be found at https://doi.org/10.1007/978-3-031-74773-1_7

Contents

1 **Introduction** 1
   1.1 General 1
   1.2 Design Philosophy for Accidental Loading 2

2 **Accidental Loading Hazard Evaluation Overview** 3
   2.1 General 3
      2.1.1 Existing Standards for Reference 4
   2.2 Hazard Evaluation Process 8
      2.2.1 Accidental Hazard Risk Assessment Plan 8
      2.2.2 Preliminary Accidental Hazard Risk Assessment 9
      2.2.3 Detailed Accidental Hazard Risk Assessment 14
      2.2.4 Documentation 15

3 **Ship Collission Hazards** 19
   3.1 General 19
      3.1.1 Existing Standards for Reference 22
   3.2 Ship Collision Evaluation 22
      3.2.1 Acceptance Criteria 22
      3.2.2 Assessment Inputs 24
      3.2.3 Ship Collision Assessment 27
      3.2.4 Mitigation Alternatives 30
      3.2.5 Documentation 31

4 **Dropped Object Hazards** 33
   4.1 General 33
      4.1.1 Existing Standards for Reference 36
   4.2 Dropped Object Evaluation 36
      4.2.1 Acceptance Criteria 37
      4.2.2 Assessment Inputs 37
      4.2.3 Dropped Object Assessment 39

		4.2.4	Mitigation Alternatives	42

5	**Fire Hazards**			43
	5.1	General		43
		5.1.1	Existing Standards for Reference	44
	5.2	Fire Evaluation		46
		5.2.1	Acceptance Criteria	47
		5.2.2	Fire Assessment Inputs	48
		5.2.3	Fire Assessment Methods	53
		5.2.4	Mitigation Alternatives	55
		5.2.5	Documentation	56
6	**Blast Hazards**			57
	6.1	General		57
		6.1.1	Existing Standards for Reference	60
	6.2	Blast Evaluation		60
		6.2.1	Acceptance Criteria	61
		6.2.2	Blast Assessment Inputs	62
		6.2.3	Blast Assessment Methods	67
		6.2.4	Mitigation Alternatives	69
		6.2.5	Documentation	69

Correction to: Accidental Load Analysis and Design for Offshore Structures ... C1

Glossary 71

References and Publications 73

Abbreviations and Acronyms

API	American Petroleum Institute
ASCE	American Society of Civil Engineers
DLB	Ductility-Level Blast
ESD	Emergency Shutdown
ETA	Event Tree Analysis
FEA	Finite Element Analysis
FMEA	Failure Modes and Effects Analysis
FPSO	Floating Production, Storage and Offloading
FTA	Fault Tree Analysis
HAZOP	Hazard Operability
HSE	Health and Safety Executive
ISO	International Organisation for Standardisation
MDOF	Multiple Degrees of Freedom
PFP	Passive Fire Protection
SCI	Steel Construction Institute
SDOF	Single Degree of Freedom
SLB	Strength-Level Blast
UFC	Unified Facilities Criteria
UKOOA	United Kingdom Offshore Operators Association (now known as Offshore Energies UK, and formerly Oil and Gas UK)
WSD	Working Stress Design

List of Figures

Fig. 2.1	Idealisation of three distinct activities in the accidental hazard evaluation process	5
Fig. 2.2	Overview of the preliminary accidental hazard risk assessment process	6
Fig. 2.3	Overview of the detailed accidental hazard risk assessment process	7
Fig. 3.1	Preliminary ship collision risk assessment process	20
Fig. 3.2	Detailed ship collision risk assessment process	21
Fig. 3.3	Idealisation of the ship collision strain energy balance. *Note* subscript 'f' is for the facility and 'v' is for the colliding vessel	28
Fig. 3.4	Three alternative approaches to predicting facility and colliding vessel strain energy. *Note* subscript 'f' is for the facility and 'v' is for the colliding vessel	29
Fig. 4.1	Dropped object preliminary risk assessment process	34
Fig. 4.2	Dropped object detailed risk assessment process	35
Fig. 5.1	Fire hazard preliminary risk assessment process	45
Fig. 5.2	Fire hazard detailed risk assessment process	46
Fig. 6.1	Blast hazard preliminary risk assessment process	58
Fig. 6.2	Blast hazard detailed risk assessment process	59
Fig. 6.3	Generic pressure curve highlighting key parameters	66

List of Tables

Table 2.1	Overview of the accidental hazard identification methods	11
Table 2.2	Overview of accidental hazard evaluation documentation requirements	16
Table 5.1	Example fire evaluation acceptance criteria	49
Table 5.2	Relationship between maximum acceptable member utilisation at ambient temperature at maximum observed member temperature [API RP 2FB]	55
Table 6.1	Example blast evaluation acceptance criteria	63

Introduction

1.1 General

This book has been developed to provide an overview of the process for identifying and assessing a variety of accidental loading scenarios that can be experienced by offshore oil and gas facilities with an emphasis on minimising health and safety, environment, and facility risks. While a wide range of potential accidents exists, the focus here is on key events that can affect the structural performance of the facility, namely ship collisions, dropped objects, fires, and blast loadings. The same conceptual process can be used with other accidental loading scenarios, such as the survival of permanent mooring systems under extreme environmental conditions. The guidance highlights activities associated with the assessment including:

(1) Accidental loading scenario hazard evaluation overview,
(2) Ship collision hazards,
(3) Dropped object hazards,
(4) Fire hazards, and
(5) Blast hazards.

This book and the guidance contained herein is not intended to serve as a design or assessment standard, but rather to highlight the primary activities relating to accidental loading assessment to promote more efficient and safer design and operation of the facility.

1.2 Design Philosophy for Accidental Loading

Accidents are defined as unintended events that arise during the course of installing, operating, or decommissioning an offshore oil and gas facility. The purpose of assessing accidental loadings is to understand the extent of initial damage and verify that the accident does not escalate in terms of personnel health and safety, environmental concerns, or facility damage (i.e., financial consequences). Escalation of the accident occurs when the local failure causes a chain of additional cascading events. For instance, a fire may impinge on a primary structural member resulting in its failure. While this may prove to be a financial consequence to the facility in terms of the member repair, an escalation of additional members failing may raise the event to include a subsequent global collapse with potential loss of life or significant environmental release.

Unlike design loadings where elastic behaviour is expected, the accidental loading assessments may consider the structure well into the plastic (inelastic) regime. The sheer magnitude of the accident makes requiring an elastic response impractical in most cases given the low likelihood of the event occurring. The acceptance criteria for the events will need to be modified to reflect the acceptability of these high strains and large deflections during the event provided that the damage is such that the event does not escalate.

The process for assessing accidental loadings follows the standard hazard risk assessment method for the oil and gas industry that incorporates traditional hazard identification methods and existing structural assessment methods. First, a preliminary risk assessment examines the potential for accidental loading based on evaluating the risk exposure that includes both the likelihood of the accident and its consequence.

A detailed accidental hazard risk assessment is then performed, if required. The events identified during the preliminary risk assessment are categorised so that low level risk and extremely low likelihood events are removed from further consideration based on the asset owner's risk tolerance. A refined assessment is then performed on the remaining accidents with the intention to either conclude sufficient structural capacity or operational constraints exist to reduce the risk (i.e., advanced analysis reduces the level of conservatism present in the initial risk assessment) or highlight mitigation activities that can successfully mitigate the risk exposure.

Accidental Loading Hazard Evaluation Overview

2.1 General

The accidental loading scenario hazard evaluation defines potential accidents that may occur to the facility during its life from installation to decommissioning and assesses the corresponding risk exposure. For the purposes of this book, an accident is defined as a scenario involving a ship collision, dropped object, fire, or blast that introduces risk to personnel, the environment, or the facility.

Fundamentally, there are two basic approaches to addressing the hazards faced by an offshore facility: hazard control and predictive hazard evaluation (API RP 14J). Hazard control requires that the facility is designed and operated in a manner consistent with Industry standard practices. While not explicitly stating as such, this approach has a built-in hazard analysis component as industry standard practices evolve to promote the safe operations of many facilities and reflect prior hazard analyses and incident investigations.

Typically, a simple checklist approach is utilised to verify standard practices are maintained from design and throughout the service life of the facility. Predictive hazard evaluation provides a method to address asset-specific hazards that may be outside of the coverage of hazard control via standards compliance. The steps involved in a predictive hazard analysis are identify pertinent hazards, evaluate the risk exposure for each hazard, and mitigate the likelihood and/or consequences associated with each event.

This book does not address the hazard control by utilising industry standard practices. It is assumed that standard practices are adhered to at all phases of the facility's design and operation. The focus on the accidental loading scenario identification study is within the execution of a predictive hazard evaluation of the proposed accidents.

The proposed accidental loading scenario evaluation process is structured in accordance with a traditional risk assessment where an initial hazard identification defines potential accidents that may occur to the facility during its life from installation to decommissioning and then the corresponding risk exposure is developed based on likelihood and consequence for each event. The process is idealised as three distinct activities: develop an accidental hazard risk assessment plan, perform a preliminary risk assessment, and perform a detailed risk assessment as warranted, as shown in Fig. 2.1.

The process begins with the definition of the hazard risk assessment method, which focuses on defining the types of accidents considered (in this case, ship collisions, dropped objects, fire, and blast), definition of the governing standards driving the evaluation, and the risk assessment guidance (including the risk matrix definition common for all hazards). Once the method is defined, the second step involves the preliminary risk assessment for specific accidents as shown in Fig. 2.2.

Accidental hazard identification defines the events including a detailed description (accident type, loading characteristics, and safeguards present) after which an assessment is performed to determine the associated likelihood and consequence from which a preliminary risk characterisation can be determined. The detailed accidental hazard risk assessment presented in Fig. 2.3 utilises information from the preliminary risk assessment to determine if the risk exposure is acceptable or if additional assessment or accident mitigations are required.

When identifying potential accidental hazards, the guidance provided in this book knowingly neglects joint events where one accident type escalates into another. For instance, a dropped object evaluation only considers the structural damage associated with the impact event and will not address the potential escalation into a fire or blast event. It is assumed that the initial hazard identification effort will highlight potential joint events for further highly specialised study where loading and structural damage reflect the joint event.

The remainder of this chapter presents a high-level process for identifying and characterising accidents and the assessment approach. Details pertaining to the execution of the hazard identification and assessment activities for a given type of accident are given in Chap. 3 through Chap. 6.

2.1.1 Existing Standards for Reference

- API RP 14J Recommended Practice for Design and Hazards Analysis for Offshore Production Facilities,
- ISO 19901-3 Petroleum and natural gas industries—Specific requirements for offshore structures—Part 3: Topsides Structure,
- API RP 2FB Recommended Practice for the Design of Offshore Facilities Against Fire and Blast Loading, and

2.1 General

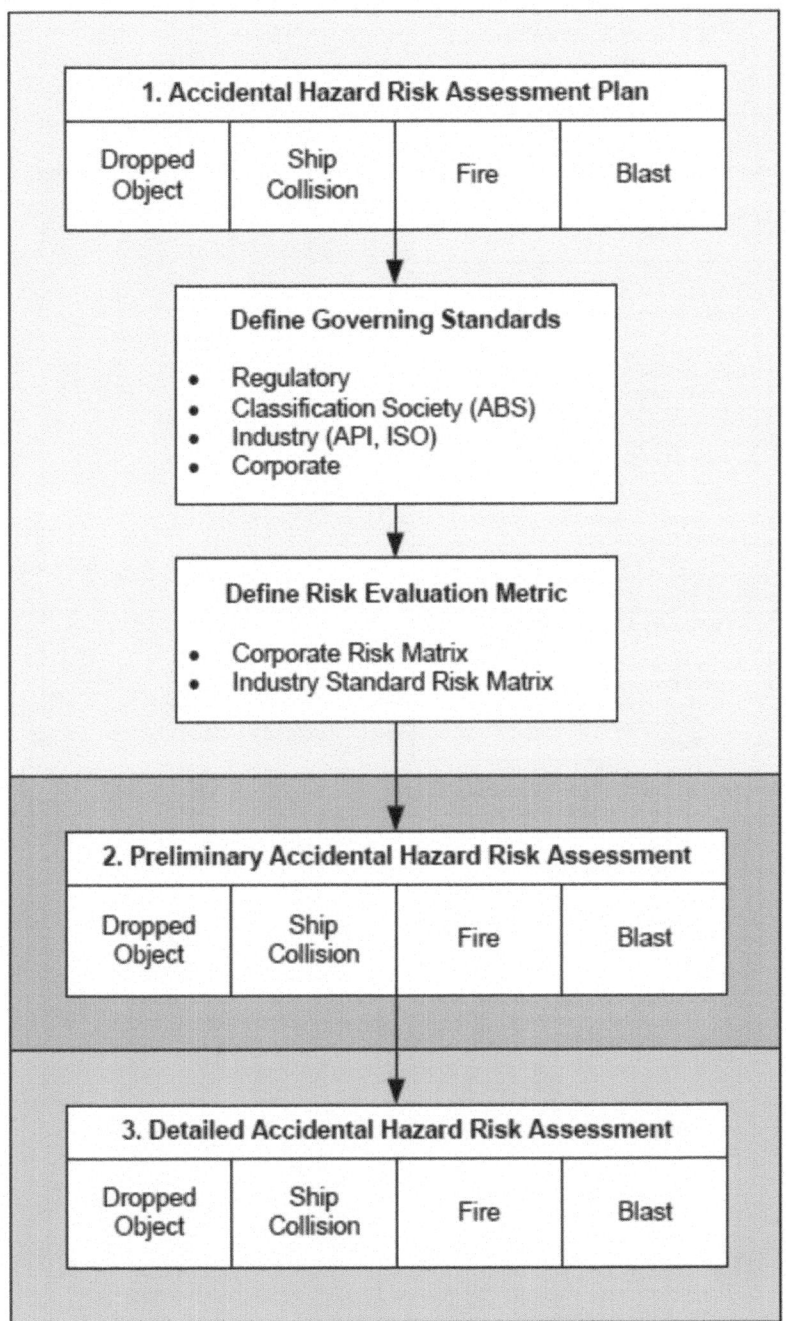

Fig. 2.1 Idealisation of three distinct activities in the accidental hazard evaluation process

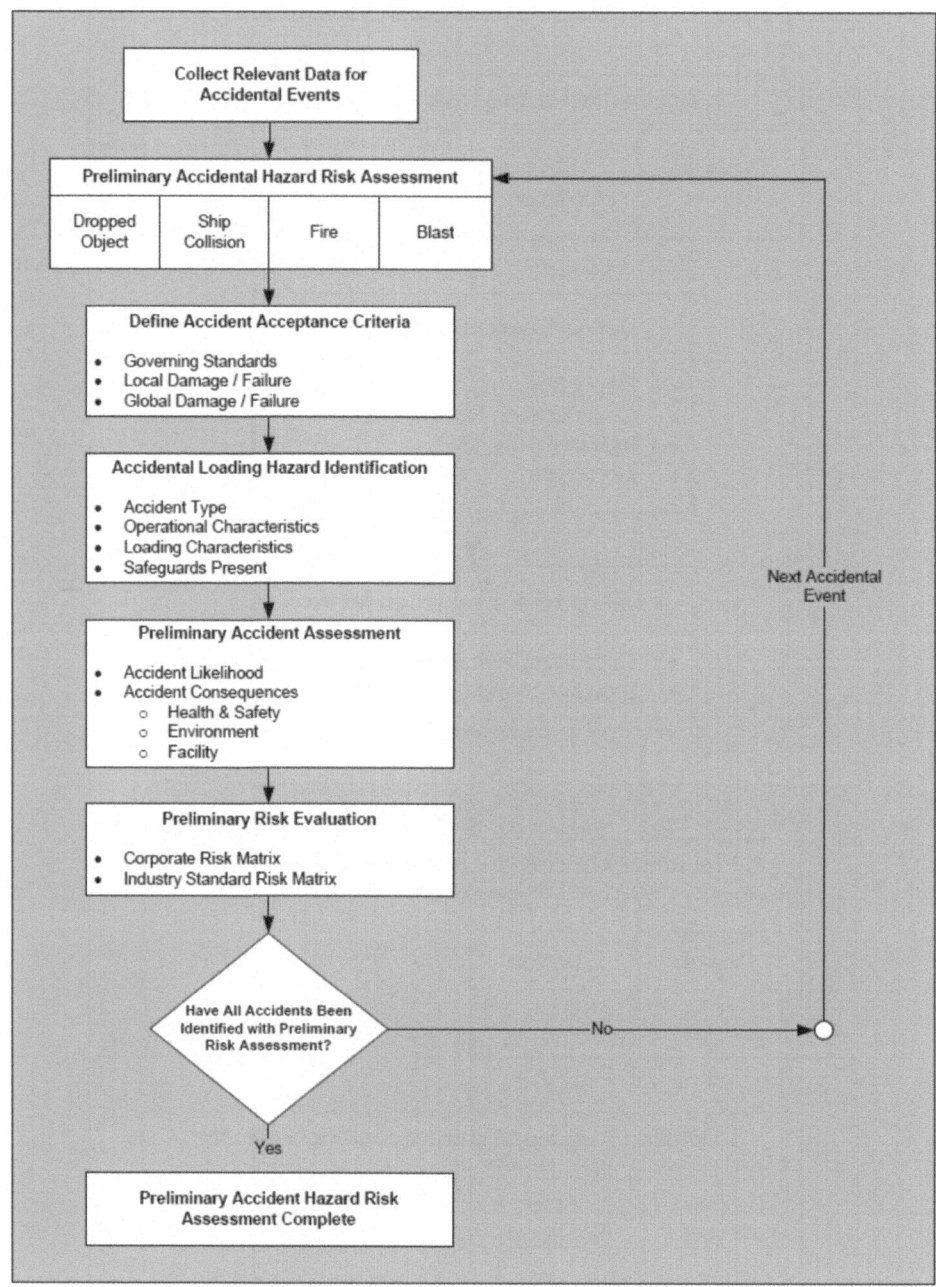

Fig. 2.2 Overview of the preliminary accidental hazard risk assessment process

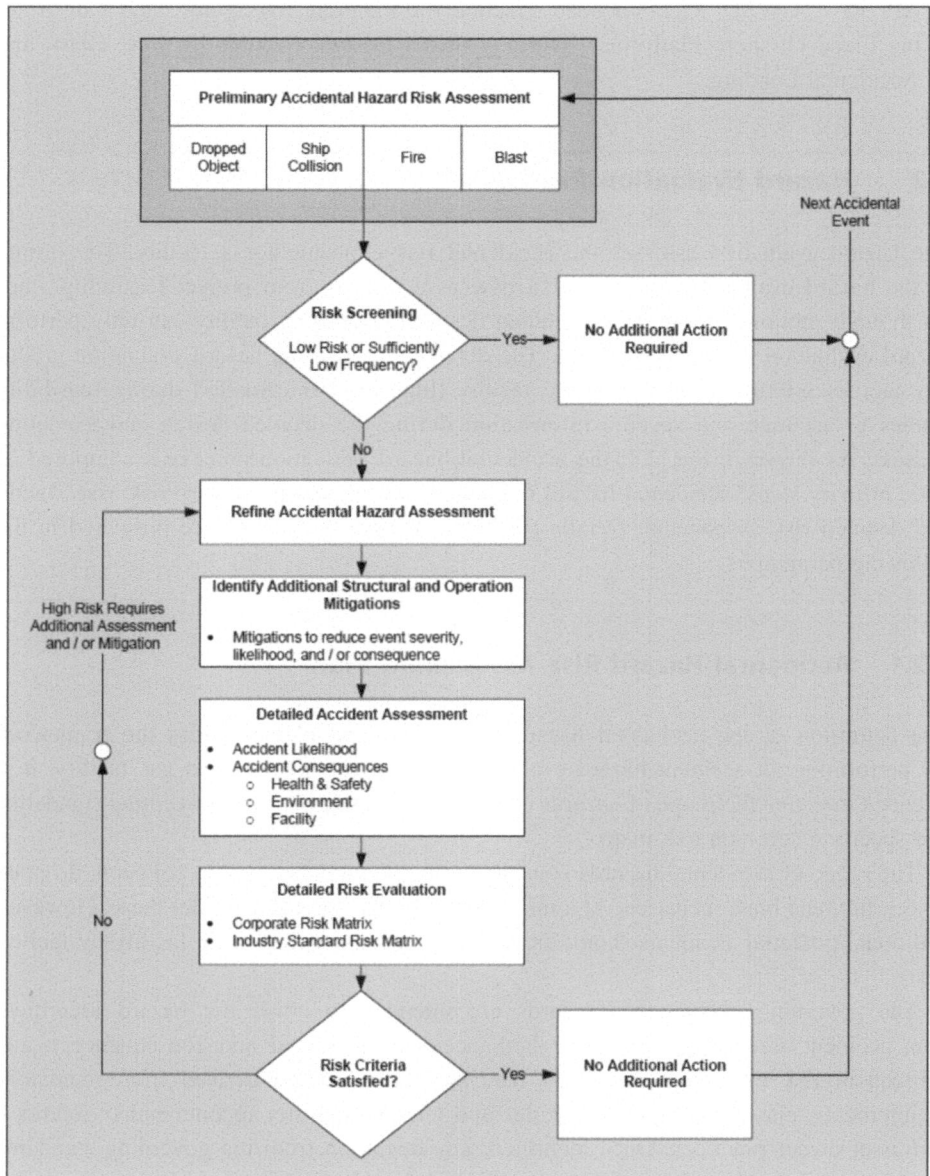

Fig. 2.3 Overview of the detailed accidental hazard risk assessment process

- API RP 2A-WSD Recommended Practice for Planning, Designing, and Constructing Fixed Offshore Platforms—Working Stress Design: Section 18, Fire, Blast, and Accidental Loading.

2.2 Hazard Evaluation Process

The hazard evaluation assesses the accidental risk exposure for a facility. The timing of the hazard evaluation process for a new asset ranges from project feasibility studies through preliminary design to detailed design (an existing facility can only perform hazard evaluation during operations). The detail present in the hazard evaluation gradually increases through the life of the facility (high-level information during feasibility studies to accurate and specific information during the detailed design and execution phases). As shown in Fig. 2.1, the accidental hazard evaluation process is composed of three primary steps: accidental hazard risk assessment plan, preliminary risk assessment, and detailed risk assessment. Details pertinent to these three steps are presented in the following paragraphs.

2.2.1 Accidental Hazard Risk Assessment Plan

The definition of the accidental hazard risk assessment plan provides the framework for performing all accident hazard evaluations for an offshore oil and gas facility. It is intended to define the scope of hazards considered, identify pertinent governing standards, and specify a common risk matrix.

The scope of accidental hazards considered by this guidance is ship collision, dropped object, fire, and blast scenarios. At a minimum, a facility should consider these. However, additional potential accidents should be addressed as appropriate on a facility-by-facility basis.

The governing evaluation standards are intended to cover the hazard identification, accident assessment (including both accident likelihood and consequences), and subsequent risk evaluation processes. The standards will be derived from regulatory requirements, classification society Rules and Guides, industry recommended practices, and asset owner practices. Once identified, any deviation from the governing standards would necessitate formal documentation highlighting the variation and its impact on the hazard evaluation.

The final piece of the risk assessment plan is in the specification of a risk matrix that will be utilised by all the evaluations. The source of the risk matrix could either be an Owner-defined matrix or a generic industry standard matrix (e.g., ISO 19901-3). Regardless of its source, the risk matrix will have at a minimum:

2.2 Hazard Evaluation Process

- Sufficient granularity in both consequence and likelihood to capture potential accidental hazard risk and differentiate between events,
- Clearly defined health and safety, environmental, and financial consequences, and
- Clearly defined risk levels with acceptability (e.g., no additional mitigation required or requires immediate mitigation prior to resuming operations).

Note that the hazard identification method and risk matrix selected for use during the hazard identification should be reviewed by Class prior to initiating the identification process if they will be requested to assess the hazard evaluation effort.

The risk evaluations done during the preliminary and detailed risk assessment phases will utilise predefined risk exposure levels. For example, three risk levels may be defined by the asset owner:

(1) *Low risk*: Insignificant/minimal risk (eliminate from further consideration),
(2) *Medium risk*: Risk can be shown to be acceptable provided additional potential mitigations are included based on cost–benefit assessment, and
(3) *High risk*: Risk requires additional mitigation to reduce the consequence or likelihood.

In addition to the three risk levels, an additional threshold for events having an "improbable" likelihood may be defined by the owner such that events with lower frequency do not require evaluation.

The definition of the evaluation method will be clearly documented so to capture the:

(1) Scope of accidental hazard evaluation,
(2) Identification of governing standards for structural analyses of each accident type and risk assessments, and
(3) Definition of the risk matrix to be used for all accidental hazard assessments.

2.2.2 Preliminary Accidental Hazard Risk Assessment

The preliminary accident risk assessment process is comprised of four key activities: definition of accident acceptance criteria, accidental loading hazard identification, preliminary accident assessment (likelihood and consequence), and preliminary risk evaluation.

2.2.2.1 Acceptance Criteria Definition

The structural performance acceptance criteria for each accident type should be defined by the owner based on existing regulatory, classification society, and corporate guidance so that the appropriate risk levels for health and safety, environment, and facility are maintained during and after the accident event. The acceptance criteria should address each potential damage mechanism faced by the structure. In general, the accidents considered will tend to induce large deformations and deflections well beyond the limits associated with traditional design standards. As such, it is incumbent on owners to clearly define the acceptance criteria so that all potential damage mechanisms as considered including:

(1) Structural member failure (primary and secondary structural members)
   - Yielding,
   - Buckling,
   - Formation of plastic hinges,
   - Excessive deformation, and
   - Connection failures (i.e., connections fail prior to member capacity).
(2) Global structural failure
   - Hull girder collapse (ship-shaped floating facility),
   - Damaged stability (fixed and floating facilities),
     – Immediately following the accident,
     – Stability until repair (e.g., capacity exceeds required return period event), and
   - Structural collapse (topsides).
(3) Safety critical elements
   - Fire/blast walls exceed design criteria,
   - Escape routes and muster areas are impinged during/after the event, and
   - Containment equipment (e.g., pressure vessels and risers) fail under accidental loading.

It is anticipated that the precise definition of the acceptance criteria will be related to the analysis method utilised to assess the accident with more robust analysis methods utilising higher acceptance criteria.

2.2.2.2 Accidental Loading Hazard Identification

The hazard identification process will be similar regardless of type of hazard considered (dropped object, collision, fire, or blast). The identification process will utilise one or more standard hazard identification methods to define the accident and its preliminary risk exposure (API RP 14J) including, but not limited to those identified in Table 2.1.

Each method or combination of methods has a place in the hazard identification process over the life of the facility, so it is important that the analysts select accordingly. As many of these methods are qualitative in nature, especially early in the design process, it is critical that a qualified and diverse staff be utilised to thoroughly detail the potential accidents.

2.2.2.3 Preliminary Accident Assessment

A preliminary accident assessment evaluates both the likelihood and consequences associated with each event identified during the hazard identification process. During this assessment, the structural response to the accident will be evaluated in light of the acceptance criteria previously defined given facility-specific information (e.g., structural

2.2 Hazard Evaluation Process

Table 2.1 Overview of the accidental hazard identification methods

Hazard identification methods	Summary of method	More common uses	Example
What-if analysis	What-if analysis is a brainstorming approach that uses loosely structured questioning to (1) postulate potential upsets that may result in mishaps or system performance problems and (2) ensure that appropriate safeguards against those problems are in place	• Generally applicable to any type of system, process or activity (especially when pertinent checklists of loss prevention requirements or best practices exist) • Most often used when the use of other more systematic methods (e.g., FMEA and HAZOP analysis) is not practical	"What happens if an offloading vessel loses station keeping capabilities during an offloading operation?" The results provide insight into subsequent activities (e.g., bow of a full offloading vessel collides with lightened FPSO at given velocity), highlighting potential mitigations
Hazard and operability (HAZOP) analysis	The HAZOP analysis technique is an inductive approach that uses a systematic process (using special guide words) for (1) postulating deviations from design intents for sections of systems and (2) ensuring that appropriate safeguards are in place to help prevent system performance problems	• Primarily used for identifying safety hazards and operability problems of continuous process systems (especially fluid and thermal systems) • Also used to review procedures and other sequential operations	A HAZOP is convened to assess risk exposure of crane operations. Attention will be paid to structural failures of the crane (from the crane pedestal to the boom) and rigging (straps/chains, shackles, and lifting points) Potential failure of each topic will be considered

(continued)

Table 2.1 (continued)

Hazard identification methods	Summary of method	More common uses	Example
Failure modes and effects analyses (FMEA)	FMEA is an inductive reasoning approach that is best suited to reviews of mechanical and electrical hardware systems The FMEA technique (1) considers how the failure modes of each system component can result in system performance problems and (2) ensures that appropriate safeguards against such problems are in place A quantitative version of FMEA is known as failure modes, effects, and criticality analysis (FMECA)	• Primarily used for reviews of mechanical and electrical systems (e.g., fire suppression systems, vessel steering/propulsion systems) • Often used to develop and optimise planned maintenance and equipment inspection plans • Sometimes used to gather information for troubleshooting systems	FMEA is performed on the gas detection system A single gas detector head is assumed to fail in the vicinity of a gas leak The analysis would consider the effect of losing a detector on the overall deluge system efficiency (i.e., volume and density of gas released prior to the impaired system detecting the release)

(continued)

2.2 Hazard Evaluation Process

Table 2.1 (continued)

Hazard identification methods	Summary of method	More common uses	Example
Fault tree analysis (FTA)	FTA is a deductive analysis technique that graphically models (using Boolean logic) how logical relationships between equipment failures, human errors and external events can combine to cause specific mishaps of interest	• Generally applicable for almost every type of analysis application, but most effectively used to address the fundamental causes of specific system failures dominated by relatively complex combinations of events • Often used for complex electronic, control or communication systems	A pool fire initiates on deck from product leaking from a crack in a cargo piping valve The FTA would consider failures contributing to the release as well as the effectiveness of the active and passive fire protection systems
Event tree analysis (ETA)	ETA is an inductive analysis technique that graphically models (using decision trees) the possible outcomes of an initiating event capable of producing a mishap of interest	• Generally applicable for almost every type of analysis application, but most effectively used to address possible outcomes of initiating events for which multiple safeguards (lines of assurance) are in place as protective features • Often used for analysis of vessel movement mishaps and propagation of fire/explosions or toxic releases	A trading vessel enters the vicinity of the facility and loses steering and/or propulsion The ETA would trace the sequence of events associated with the stricken vessel striking the facility

configuration and material properties). This response will in turn provide direct input into the specification of both the likelihood and consequences for each event.

2.2.2.4 Preliminary Risk Evaluation

A preliminary risk is evaluated for each event based on the likelihood and consequences developed during the accident assessment phase. The resulting risk exposure will identify key events to be examined in subsequent detailed assessment. Ideally for consistency, the same risk matrix should be utilised for all accidental loading scenarios as well as other design and operational aspects of the facility.

2.2.2.5 Preliminary Risk Evaluation

The output for the preliminary accidental hazard risk assessment process includes:

(1) Detailed descriptions of and basis for each accident including event type, location, and intensity (e.g., type of vessel anticipated for a ship collision, energy of dropped object, flame geometry and duration, and blast overpressure and duration at different locations),
(2) Safeguards present (e.g., design margins, operations, and active/passive safeguards present),
(3) Acceptance criteria assumed during evaluation, and
(4) Preliminary likelihood, consequence, and risk evaluation.

2.2.3 Detailed Accidental Hazard Risk Assessment

The detailed accidental hazard risk assessment considers two primary steps: an initial event screening and a refined hazard assessment. The progression of the detailed accidental hazard risk assessment shown in Fig. 2.3 relates activities to the anticipated risk and consequence levels for a given event (note that this assumes the Owner specifies low, medium, and high-risk levels during risk matrix definition):

(1) If low risk is present or if the event frequency is sufficiently low, then no additional action is required,
(2) Additional detailed hazard assessment should be performed on the high and medium risk events to either reduce the predicted risk via more refined (less conservative) modelling or identify additional mitigators,
(3) If medium risk is obtained using the refined assessment and all reasonable safety barriers are present (i.e., all mitigations are present except those identified as being grossly disproportionate in expense, time, or effort relative to the reduction in risk they afford), then no additional action is required, and
(4) If high risk is present, then additional detailed assessment and mitigation evaluations should be pursued until medium or low risk is obtained.

Once the risk for the current event has been reduced to acceptable limits, the process is repeated with the next event from the hazard identification.

The specific details regarding the detailed hazard assessment procedures for each of the key accidents of ship collision, dropped objects, fire, and blast are detailed in Chap. 3 through Chap. 6, respectively. The detailed accidental hazard risk assessment will be documented in a manner similar to that with the preliminary assessment. The primary distinctions are that the detailed assessment will not include specifics on the hazard identifications but may include multiple analyses of each accident.

2.2.4 Documentation

The documentation requirements for the accidental hazard evaluation are highlighted in Table 2.2. Three activities are defined:

(1) Accidental hazard risk assessment plan,
(2) Preliminary accidental hazard risk assessment, and
(3) Detailed accidental hazard risk assessment.

Where Class is requested to review the accidental hazard evaluation, each of these activities should be submitted to Class prior to proceeding to subsequent activities. For activities 2 and 3 (preliminary and detailed accidental hazard risk assessments, respectively), it is assumed that the individual hazard types (ship collision, dropped objects, fire, and blast) will be evaluated separately. In this case, the detailed risk assessment for a given hazard type (e.g., fire) only requires that its preliminary risk assessment has been reviewed by Class. It should be noted that the elements presented in the preliminary and detailed accidental hazard risk assessment deliverables (2 and 3) are generic. Details pertaining to a specific hazard identification and assessment deliverables are provided in Chap. 3 through Chap. 6.

Table 2.2 Overview of accidental hazard evaluation documentation requirements

Description	Overview of the method to evaluate potential accidental loadings pertinent to the facility	
Includes	Examples	Detailed in [a]
1. Accidental hazard risk assessment plan		
Hazards to be considered	Ship collisions, dropped objects, fire, and blast events	2/1
Pertinent governing standards for each hazard type	Class standards, industry standards (API, ISO), or asset owner standards	X/1.1
Risk matrix to be utilised during evaluation	Industry standards (API, ISO) or asset owner standard	2/2.1
2. Preliminary accidental hazard risk assessment[a]		
Hazard identification process utilised for each hazard type	Hazard operability analysis, failure modes and effects analysis, and fault tree analysis	2/2.2
Pertinent governing standards for hazard type	Class standards, industry standards (API, ISO), or asset owner standards	X/1.1
Acceptance criteria definition	Definition of local acceptance criteria (e.g., deflection and stresses) and global acceptance criteria (e.g., collapse and facility stability)	X/2.1
Personnel involved in the identification process	Personnel with diverse experience capable of identifying potential accidents and their likelihood and consequence	2/2.2.3
Accidental event definition (magnitude, location, and safeguards present)	Ship collision would include the vessel type (supply operations or drifting vessel), facility and colliding vessel characteristics (mass, hull shape/form, safety barriers), collision specifications (velocity, approach angle, contact location, environmental conditions, and applied loading	X/2.2
Preliminary risk evaluation based on likelihood and consequence	Based on event definition and acceptance criteria, a preliminary risk assessment is developed based on the experiences and judgment of the hazard identification personnel (most likely qualitative risk assessment)	2/2.1
3. Detailed Accidental Hazard Risk Assessment[a]		
Description	*Accidental hazard assessment process for each hazard type considered*	
Acceptance criteria definition	Definition of local acceptance criteria (e.g., deflection and stresses) and global acceptance criteria (e.g., collapse and facility stability)	X/2.1

(continued)

2.2 Hazard Evaluation Process

Table 2.2 (continued)

Description	Overview of the method to evaluate potential accidental loadings pertinent to the facility	
Includes	Examples	Detailed in [a]
Detailed description of the accidental loading	Reference that developed during the hazard identification activity	X/2.2
Pertinent governing standards for hazard type	Class standards, industry standards (API, ISO), or asset owner standards	X/1.1
Assessment method and results	Assessment method may involve sequentially more complicated assessment processes to reduce the conservatisms present A fire assessment may initiate with a zone method (screening analysis) then proceed to a linear-elastic method (strength-level analysis) and nonlinear, elastic–plastic method (ductility-level analysis) Results presented would, at a minimum, provide sufficient information to address the selected acceptance criteria	X/2.3
Proposed additional structural and operational mitigations (if required)	Dropped object mitigations may include adjusting the general arrangement of critical equipment, restricting crane operations, and increased structural protection	X/2.4
Detailed risk exposure	Risk assessment process utilising updated likelihood and consequence (refined assessment and additional mitigations)	2/2.1

[a] Reference nomenclature is "X/Y.Z" where X is the chapter for each hazard and Y.Z is corresponding section or paragraph. For brevity, the use of "X" references discussion on ship collision (3), dropped objects (4), fire (5), and blast (6). If only a single section is appropriate, "X" is replaced by a chapter number

Ship Collision Hazards 3

3.1 General

The ship collision assessment provides an overview to the evaluation of potential impact events between the offshore facility and nearby vessels that could lead to significant facility damage and even escalate into personnel life/safety incidents or environmental releases. The method presented is applicable to both new and existing facilities. However, the ability to implement additional measures to mitigate damage potential for existing facilities may be significantly restricted.

The ship collision event typically begins with a nearby support, off-loading, or trading vessel losing its station keeping ability due to mechanical, electrical, or environmental reasons. The vessel then strikes the fixed or floating facility causing damage to both structures. The damage is typically associated with large structural deformations resulting in the potential for dented tubular members and cracked joints for fixed platforms or distorted and ruptured shell plating and stiffeners for floating facilities.

The general process is as follows:

(1) Ship collision preliminary risk assessment (Fig. 3.1)
  - Define acceptance criteria,
  - Define collision events,
  - Perform collision assessment, and
  - Perform preliminary risk evaluation.
(2) Ship collision detailed risk assessment (Fig. 3.2)
  - Select pertinent/applicable collision scenarios from ship collision preliminary risk assessment,
  - Identify mitigation actions to meet assessment criteria or re-assess the event using a more advanced (less conservative) analysis method,

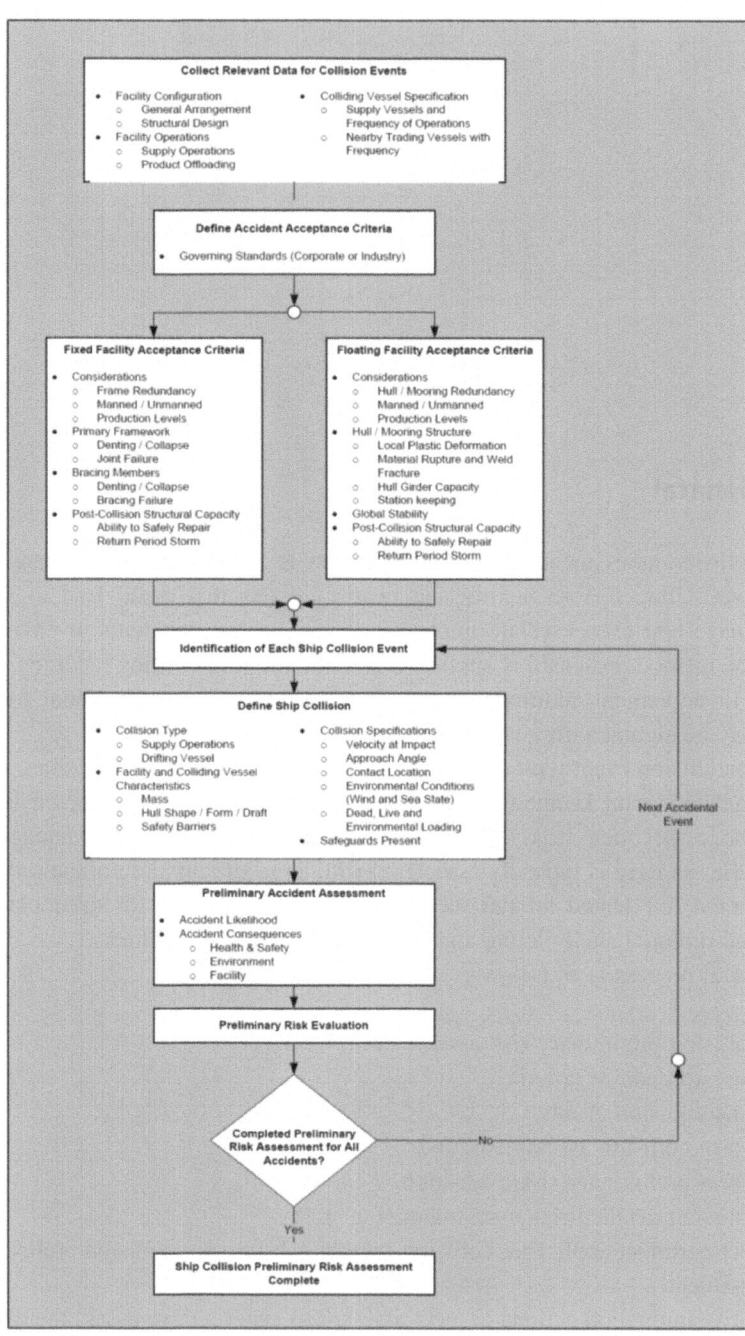

Fig. 3.1 Preliminary ship collision risk assessment process

3.1 General

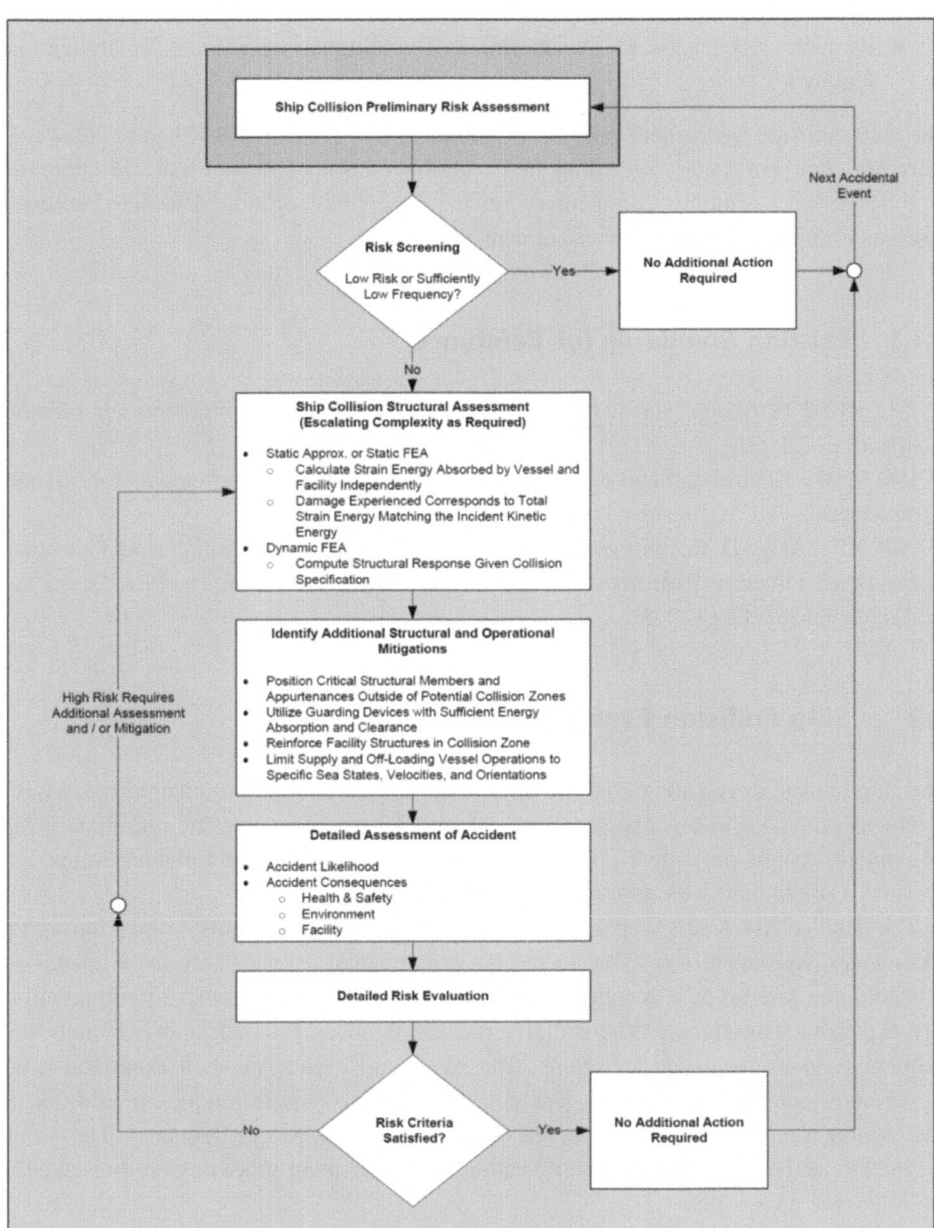

Fig. 3.2 Detailed ship collision risk assessment process

- Evaluate global and local performance, and
- Evaluate risk exposure to determine if additional assessment or mitigations required.

The ship collision assessment process is similar to that presented in Chap. 2. However, there are some particulars regarding its execution for ship collision that are addressed in this Section, including commentary on the acceptance criteria definition, collision assessment inputs, and collision assessment.

3.1.1 Existing Standards for Reference

- ISO 19902 Petroleum and natural gas industries—Specific requirements for offshore structures [2007],
- ISO 19901-3 Petroleum and natural gas industries—Specific requirements for offshore structures—Part 3: Topsides structure [2010], and
- API RP 2A-WSD Recommended Practice for Planning, Designing, and Constructing Fixed Offshore Platforms—Working Stress Design: Section 18, Fire, Blast, and Accidental Loading [2008].

3.2 Ship Collision Evaluation

The ship collision evaluation consists of two steps: preliminary and detailed risk assessments, as presented in this chapter, Figs. 3.1 and 3.2, respectively. The preliminary risk assessment step, as presented in Chap. 2, defines the potential ship collision events, and provides a preliminary risk assessment for each accident.

The detailed risk assessment step initiates with the collision events defined during the preliminary assessment step. Due to the large number of events likely to be identified, it is common to employ a screening approach to minimise the number of computationally expensive assessments. The initial screening removes low risk and extremely low likelihood events from considerations. The remaining events are then examined using increasingly complex (and hence, less conservative) analysis techniques in addition to determining if mitigation alternatives are viable to reduce the risk or frequency. The details for the key activities within the identification and assessment processes are presented in the following paragraphs.

3.2.1 Acceptance Criteria

The collision acceptance criteria define structural response thresholds that will be utilised during the risk assessment. Acceptance criteria for all affected members and systems

3.2 Ship Collision Evaluation

should be defined by the asset owner so that the health and safety, environment, and facility risk levels are properly characterised during and after the collision event. The evaluation process considers the potential failure of individual structural elements (e.g., a horizontal brace in a fixed platform) and the global performance (e.g., maintaining structural stability after rupturing the side shell of a floating production facility). In addition to the structural concerns, the assessment should also verify that all applicable safety critical elements are not adversely affected by the ship collision event. It is assumed that in addition to satisfy the aforementioned asset owner specified acceptance criteria, the criteria will satisfy all pertinent regulatory requirements such as:

- Norwegian Maritime Directorate Regulations of 10 February 1994 no. 123 for Mobile Offshore Units with Production Plants and Equipment, as referenced in ABS *Guide for Mobile Offshore Units Operating on Norwegian Continental Shelf, N-Notation*, February 2012, identifies collision with a 5,000 ton displacement supply vessel at a speed of 2 m/s, and
- Canada Oil and Gas Installations Regulations (SOR/96-118) states that every platform shall be capable of absorbing the impact energy of not less than 4 MJ from a vessel without endangering any person or the environment.

During the collision event, the kinetic energy of the striking vessel will be converted into strain energy and motion of the involved ship and offshore structure (fixed or floating) during the event. As a result, the primary structural effects associated with a collision event relate to the plastic deformation and failure of structural members involved in the collision. The associated damage mechanisms can be significantly different between a fixed and floating offshore facility due primarily to the difference in structural members utilised (tubular versus stiffened shells) and redundancy present.

A floating facility has a variety of potential damage mechanisms depending on the specifics associated with the collision event. For example, when considering an FPSO subjected to collision, the following performance measures are typically considered:

- Release of hydrocarbons to the environment (e.g., risers and conductors, ruptured cargo oil tanks),
- Structural failures (primarily plastic deformations, material rupture, and weld failures),
- Damage stability (in operational environment), and
- Hull girder capacity.

For lower energy collisions, the consequences are primarily governed by the financial cost to repair the local damage (including potential lost or deferred production revenue). Fixed platforms typically have less redundancy than a floating facility and require slightly different acceptance criteria. It is assumed that portions of the structure will plastically deform during the collision event in order to absorb the energy. Acceptance criteria should define collision energy absorption design requirements for the primary framework (bracing and jacket legs) as well as risers and conductors. The energy absorption of tubular members can be divided into two primary components: local denting and global bending.

A variety of solutions exist for the local denting force of tubular members. While the format may vary slightly, the relationship between dent depth and applied force is a function of material yield strength and the tubular geometry (tubular diameter and wall thickness). Integrating this denting force over the dent depth provides the energy absorbed that can be directly related to the collision kinetic energy. In addition to loading event considerations, there are additional post-event considerations required for the fixed platform analogous to the stability of floating structure, namely the facility should be designed so that the damaged structure can resist a one-year environmental storm or current load in addition to normal operating loads (API RP 2A-WSD). The assessment of the damaged facility should include reasonable representation of actual stiffness of damaged member or joints.

All acceptance criteria and corresponding risk shall be defined and documented by the asset owner.

3.2.2 Assessment Inputs

The collision assessment is composed of four primary inputs:

(1) Collision scenario definition,
(2) Structural configuration,
(3) Material properties, and
(4) Applied loading.

A brief overview of each input is presented in the following paragraphs.

3.2.2.1 Collision Scenario Definition

The collision scenarios are identified using the process outlined in Chap. 2. The scenarios considered will include collision events such as:

- Regularly visiting supply vessel collision,
- Passing vessel collision, and
- Off-loading shuttle tanker collision.

The relative size, velocity, and frequency of collision events will be based on best available information for the site. It will include information such as the size and velocity of anticipated (or actual) supply vessels, vessels in shipping lanes, and off-loading shuttle tankers. For instance, the supply vessel size is significantly different depending on operational region (ISO 19902, API RP 2A-WSD):

- North Sea (northern): Vessel mass of 8,000 tons,
- North Sea (southern): Vessel mass of 2,500 tons, and
- Gulf of Mexico (shallow water): 1,000 tons (represent typical 55–60 m supply vessel; larger vessels are typically assumed for deepwater facilities and smaller for shallower).

Similar site-specific vessel size variation for trading vessels and off-loading tankers would be anticipated as well. The anticipated velocity of the incident vessel will vary based on the scenario. It may be as low as 0.5 m/s for a minor (low energy) collision with a supply vessel or 2 m/s for a drifting supply vessel (ISO 19902). Due consideration is to be given to the effects of environmental conditions on the velocity at impact (i.e., current, wind, and sea state). While historical information may be available, it is incumbent on the analyst to estimate the current and future (via trending) ship sizes and frequency of operations around the platform.

It is common for the collision analysis to consider a variety of potential events. Typically, these are divided into low-energy and high-energy collisions. The low-energy events will typically result in structural damage not requiring major repairs, remediation, or facility shut down. The high-energy events are lower likelihood events that severely damage the hull of a floating facility or the fixed platform jacket. When selecting the scenarios to consider in more detail, it is assumed that a minimum likelihood considered will be utilised consistent with the structural design and accidental loadings.

The collision event definitions will be fully characterised including:

(1) On-going operations (i.e., supply, offloading, or drifting trading vessel)
(2) Incident vessel description:
   - Mass, and
   - Characteristic properties (e.g., hull shape, form, draft, etc.).
(3) Collision definition:
   - Velocity at impact,
   - Approach angle,
   - Location of contact (including floating facility draft if appropriate),
   - Environmental conditions (i.e., current, wind, and sea state) during event and for evaluating post-event stability,
   - Safety measures present (e.g., fenders or riser guards), and
   - Likelihood of event.

3.2.2.2 Structural Configuration

The structural geometry considers the structures for both the facility and the impacting vessel. Sufficient information is required for both the facility and impacting vessel to develop global and local structural models to assess the collision event. A local structural model will be developed at the collision point that is capable of capturing the high deformations, yielding and failure of adjacent structural members for both the offshore facility and the impacting vessel. This local model will either utilise a sufficiently large portion of the structures so that boundary conditions become irrelevant to the local response or employ realistic boundary conditions taken from a global model.

Structural definition should include particular attention to structures utilised as part of or influencing safety barriers (e.g., proximity to critical safety elements, personnel evacuation routes/muster areas, and degree to which structural damage can be tolerated). A global model is also developed to assess the damaged structure's overall in-place performance prior to enacting permanent repair or mitigation.

In additional to the aforementioned structural models, it may be necessary to characterise the hydrodynamics of the floating facility and impacting vessel so that vessel motions and stability are properly captured during and after the collision event.

3.2.2.3 Material Properties

The stiffness, strength, and ductility of individual structural components are the primary material properties associated with a collision assessment. It may be appropriate to incorporate strain-rate effects into the assessment based on collision loading rate.

3.2.2.4 Applied Loading

The applied loads for a collision assessment address the various structural loading components present, namely:

- Dead loads,
- Live loads,
- Environmental loads, and
- Collision loads.

Dead loads are composed of structural member and equipment weights. Live loads reflect the variable loads associated with the operation of the facility. The live load magnitude selected should be consistent with the other design and accidental loading assessments performed. The final piece of the applied loading that defines the stress state in the structure prior to the collision event is the environmental loading. It is important that the environmental loads reflect those defined during the hazard assessment. Environmental loading is primarily concerned with the facility stability subsequent to the collision event (i.e., the reduction in structural capacity does not cripple the fixed platform or reduce the stability margins for a floating facility to unacceptable levels).

Definition of the collision load inputs will be defined via the initial hazard identification and assessment process described in Chap. 2. While the initial loading definition will be the same for the collision assessment (a kinetic energy component), the evolution of the applied collision load with respect to time will have a significant variation based on modelling assumptions (i.e., stiffness variation will change contact area which will alter deformations/strains and ultimately affect the strain and kinetic energy components).

3.2.3 Ship Collision Assessment

The ship collision assessment problem can be separated into two distinct mechanisms: external and internal mechanics. External mechanics addresses the motions of the vessel and facility involved in the collision. This study develops the time-dependent rigid body motions and accounts for the accidental and hydrodynamic forces. The internal mechanics focus on the structural failures of the bodies. By ensuring conservation of energy, it is possible to equate the strain energy developed in the impacting vessel and facility (energy absorbed) to the loss in kinetic energy. The collision damage assessment assumptions pertaining to the handling of the internal and external collision mechanics will have a direct effect on the subsequent analysis complexity and the resulting conservatism. The ship collision energy balance can be expressed as:

$$E_{KE,int} = E_{ext} + E_{int} \tag{3.1}$$

where the initial kinetic energy of the colliding vessel ($E_{KE,int}$, initial colliding vessel kinetic energy) is converted into energy via both external mechanics (E_{ext}) and internal mechanics (E_{int}). Equation (3.1) can be decomposed into the facility and colliding vessel contributions and re-arranged to highlight the relationship between the change in kinetic energy and developed strain energy:

$$E_{KE,int} - \left(E_{KE,f} + E_{KE,v}\right) = \left(E_{SE,f} + E_{SE,v}\right) \tag{3.2}$$

where:
$E_{KE,f}$ = facility kinetic energy after contact.
$E_{KE,v}$ = colliding vessel kinetic energy after contact.
$E_{SE,f}$ = facility strain energy.
$E_{SE,v}$ = colliding vessel strain energy.

A variety of assumptions can be utilised that will simplify Eq. (3.2). For example, neglecting external mechanics means there is no kinetic energy after contact so that the strain energy is equivalent to the initial kinetic energy. A variety of approaches have been developed to account for the change in kinetic energy resulting from colliding vessel and facility surge, sway, and yaw motions (in general, the other three motion components can be neglected for most collision problems). Depending on the model complexity, inputs include not only the initial collision velocity and mass of the colliding vessel but also its heading, the collision location, and hydrodynamic properties of the vessel and facility.

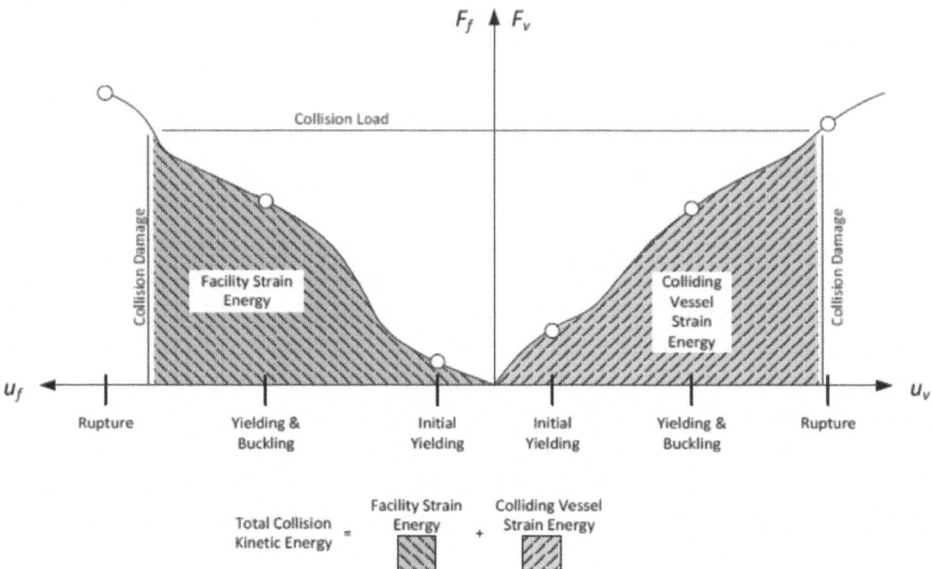

Fig. 3.3 Idealisation of the ship collision strain energy balance. *Note* subscript 'f' is for the facility and 'v' is for the colliding vessel

During the collision, strain energy is developed via structural deformation. The resulting collision event energy balance can be visualised as shown in Fig. 3.3. The visualisation considers the applied collision load (F) to both the facility and colliding vessel and the resulting deformation (u). As the deformation increases, the damage present also increases from local buckling early on in the collision process to rupture of the outer shell and ultimately, if the applied load is high enough, rupture of the inner shell as well.

Recognising from Eq. (3.2) that the change in kinetic energy is equivalent to strain and kinetic energy, and that the same collision load is applied to the facility and vessel, it is possible to determine the damage present. During the process of setting up collision scenarios (load cases) for analyses, it is important to consider the possible collision directions, and other companion loads if any. During the analysis, the uncertainty in ship geometry should be addressed. If the analysis is to be performed via dynamic finite element method (discussed below), the uncertainty of impacting speed should also be addressed.

There are three methods to determine the strain energy (and hence, damage present) in the facility and colliding vessel in Fig. 3.4:

(1) *Method A:* Rigid body colliding with the facility A_f,
(2) *Method B:* Two independent analyses considering a rigid body colliding with the facility Bf and a deformable colliding vessel impacting a rigid facility B_v, and

3.2 Ship Collision Evaluation

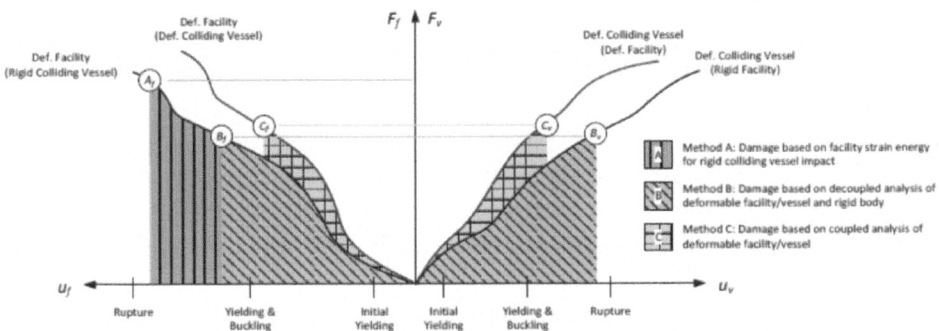

Fig. 3.4 Three alternative approaches to predicting facility and colliding vessel strain energy. *Note* subscript 'f' is for the facility and 'v' is for the colliding vessel

(3) *Method C:* Coupled analysis of a deformable facility C_f and colliding vessel C_v.

In *Method A*, it is assumed that the entire change in kinetic energy is accounted for by facility deformation. This is the most conservative of the three alternatives and will indicate the most severe facility damage.

Method B considers two separate analyses where a rigid body impacts the facility and colliding vessel. This is less conservative than the first approach as it includes the strain energy from the colliding vessel with the facility. While the damage curve for the facility is unchanged from the first approach, the total applied load may be significantly less.

Method C is the least conservative of the three, but the most computationally expensive, as it not only accounts for strain energy developed in both bodies but also considers the effect of impact force being spread out over a larger area as the two bodies deform.

There are two broad categories of methods to address the internal mechanics associated with the three approaches for predicting the strain energy of the ship collision event:

(1) Static approximate or finite element analysis, and
(2) Dynamic finite element analysis.

These are described in the following paragraphs.

3.2.3.1 Static Approximate or Finite Element Analysis

The quasi-static analysis method relates the change in kinetic energy to the structural strain energy dissipation of the striking vessel and the facility. The primary step in the quasi-static analysis is to develop the strain energy stored in both the striking vessel and facility such that it equals the change in kinetic energy due to the collision. To do this, it is necessary to evaluate the energy dissipated via elastic and plastic deformations of the two structures. These deformations are complex and will typically involve large plastic strains and potentially structural failures (e.g., plate/stiffener rupture). In an effort to simplify the assessment, the strain energies for the two bodies may be addressed separately from one another (i.e., damaged based on decoupled analyses with facility/vessel impacted by rigid body as shown by *Method B* in Fig. 3.4).

(1) For typical supply vessels and installations built from regular steel shapes, tables and figures may be available from references (such as API RP 2A-WSD) that can be used to evaluate the strain energy stored in both the striking vessel and impacted facility, and
(2) To develop force–deformation curves to be used for the evaluation of strain energy, nonlinear finite element analysis is often performed based on given structural configuration, material properties and collision scenario. Simplified plastic analysis techniques are sometimes also employed for this purpose.

3.2.3.2 Dynamic Finite Element Analysis

Nonlinear finite element analyses provide a means to more accurately assess the internal mechanics of the collision event while providing details on virtually every structural aspect of the event. Properly modelled nonlinear, dynamic finite element simulations can capture many of the failure mechanisms anticipated for a ship collision event by consideration of the subsequent stress state during the event. The structural failure modes may include material yielding, large deformations of plate and stiffener components, connection failures, and material rupture. The modelling of the collision event typically requires a highly refined model in the vicinity of the contact area to capture not only the structural deformations but also the changing contact area between the bodies and the desired failure modes. The dynamic simulation will then trace the event from its initial contact through to the termination of the striking vessel's velocity (assumption during development was that the external mechanics required all the velocity to be absorbed as strain energy).

3.2.4 Mitigation Alternatives

Mitigation alternatives typically focus on reinforcing the facility's structure exposed to the collision and modifying acceptable operational practices, including but not limited to:

- Locating fixed platform structural elements carrying dead loads (aside from legs and piles) outside of potential collision zones,
- Utilising guarding devices with sufficient energy absorption and clearance to protect exposed facility appurtenances including risers and conductors (*Note: care must be taken when designing the guarding device backup structure*),
- Specifying materials and welds with sufficient toughness to resist rupture/fracture,
- Using clear warnings to identify unprotected appurtenances,

- Reinforcing facility structures in collision zone, and
- Limiting supply and off-loading vessel operations to specific locations around the facility, sea states, speeds, and orientations.

3.2.5 Documentation

Documentation for the ship collision hazard evaluation process will address the following:

(1) Preliminary ship collision risk assessment (Fig. 3.1), and
(2) Detailed ship collision risk assessment (Fig. 3.2).

The preliminary ship collision risk assessment documentation provides a brief summary of the activities in Fig. 3.1:

- Acceptance criteria for structural and safety critical elements,
- Identified collision scenarios considered, and
- Preliminary risk assessment.

Where Class is requested to review the ship collision hazard evaluation process, it is prudent to provide the preliminary risk assessment to Class for review prior to initiating additional collision assessment activities. The detailed ship collision assessment documents pertinent inputs and key structural outputs as presented in Fig. 3.2:

(1) Discussion of screening process for collision events requiring additional assessment,
(2) Description of the structural response of the collision with subsequent stability as required by acceptance criteria,
(3) Identification of all required mitigation actions for safe and efficient operations, and
(4) Detailed risk assessment.

Dropped Object Hazards

4.1 General

Dropped objects encompass a wide variety of accidents from objects dropped during lifting operations from a supply boat or drilling operations to unsecured debris falling overboard during a storm. In each case, a dropped object develops a significant amount of kinetic energy that may significantly impair the performance of structures and equipment impacted. The focus of the dropped object evaluation process will be on the effect of a dropped object on topsides structures and equipment as well as near water surface locations where the trajectory of the falling object is considered to be unaffected by the sea water, wave action, and currents. The primary damage typically observed is dented or ruptured members, depending on the dropped object's mass, shape, and velocity.

While the risk exposure to deeper water structures (either fixed jackets or hull systems) and subsea equipment may have the potential for significant facility, health and safety, or environmental release, this class of dropped objects will not be addressed herein. Proper study of these events requires specialised techniques to address the dropped object trajectory and subsequent likelihood of striking additional structure and equipment as well as predicting the consequences of such subsequent impacts.

The general evaluation process for dropped object accidents is as follows:

(1) Dropped object preliminary risk assessment (Fig. 4.1):
- Define acceptance criteria,
- Define dropped object events,
- Perform dropped object assessment, and
- Perform preliminary risk evaluation.

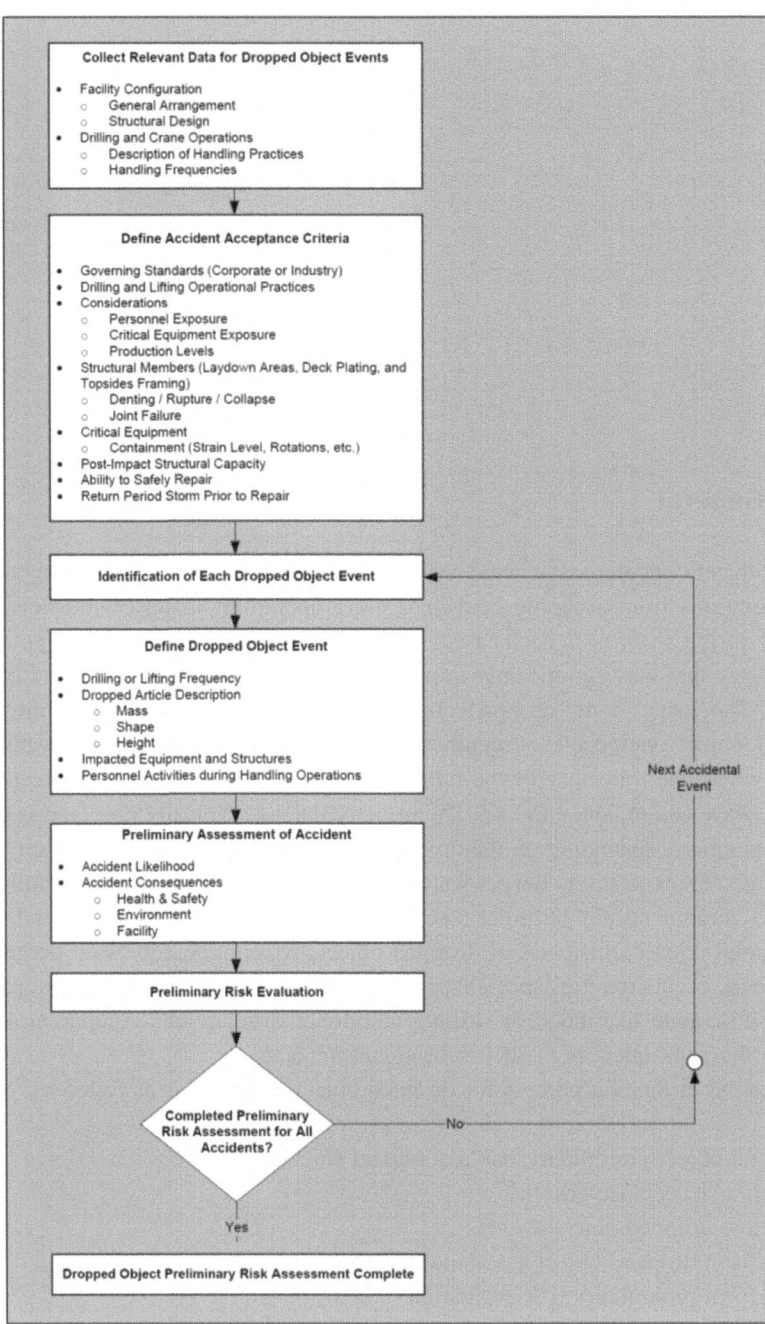

Fig. 4.1 Dropped object preliminary risk assessment process

4.1 General

(2) Dropped object detailed risk assessment (Fig. 4.2):
- Select pertinent/applicable scenarios from the dropped object preliminary risk assessment,

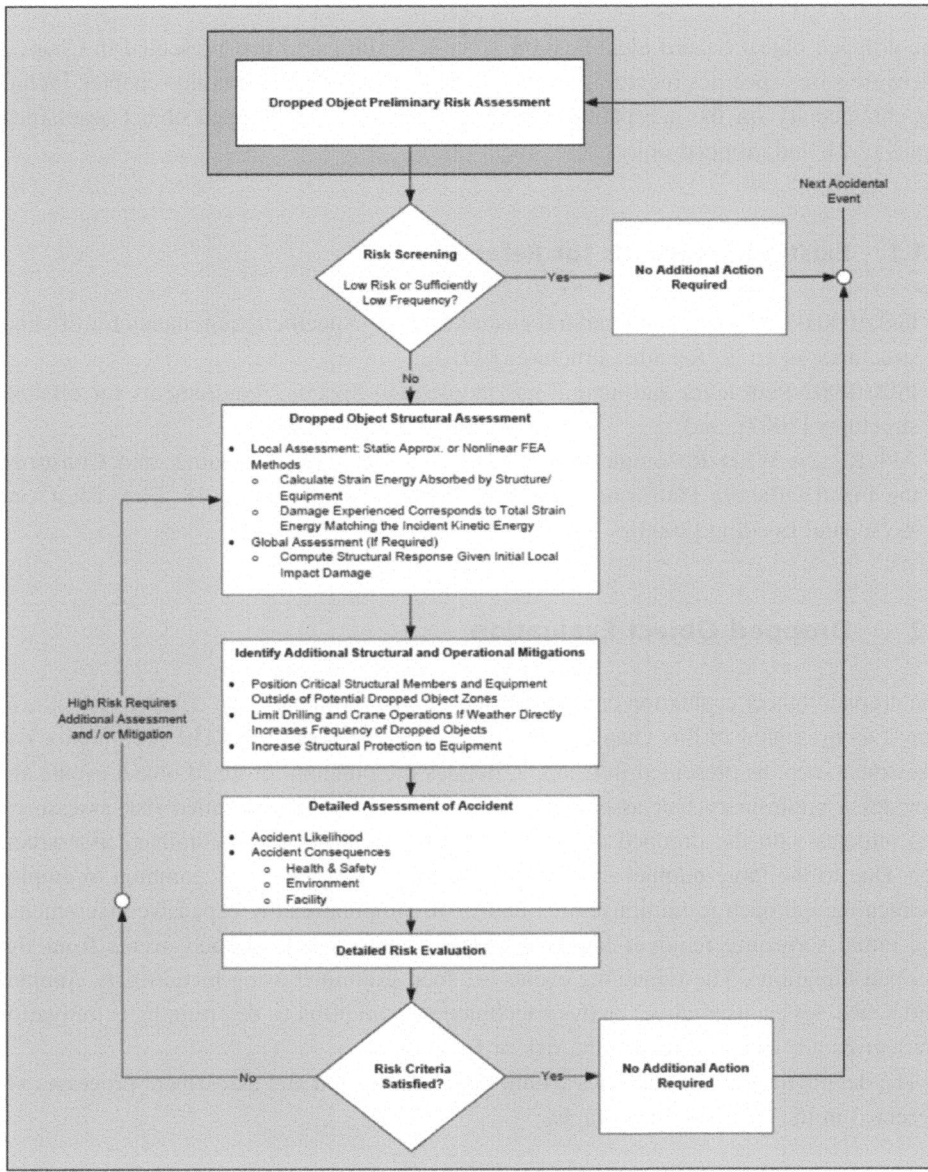

Fig. 4.2 Dropped object detailed risk assessment process

- Identify mitigation actions to meet assessment criteria or re-assess the event using a more advanced (less conservative) analysis method,
- Evaluate local and global performance, and
- Evaluate risk exposure to determine if additional assessment or mitigations required.

The dropped object hazard identification process is similar to that presented in Chap. 2. There are some specifics regarding its execution that are addressed in this chapter, including commentary on the acceptance criteria definition (2.1), dropped object assessment inputs (2.2), and dropped object assessment (2.3).

4.1.1 Existing Standards for Reference

- ISO 19901-3 Petroleum and natural gas industries—Specific requirements for offshore structures—Part 3: Topsides structures [2010],
- ISO 19902 Petroleum and natural gas industries—Specific requirements for offshore structures [2007], and
- API RP 2A-WSD Recommended Practice for Planning, Designing, and Constructing Fixed Offshore Platforms—Working Stress Design: Section 18, Fire, Blast, and Accidental Loading [2008].

4.2 Dropped Object Evaluation

The dropped object evaluation consists of two steps: preliminary and detailed risk assessments, as presented in this chapter, Figs. 4.1 and 4.2, respectively. The preliminary risk assessment step, as presented in Chap. 2, defines the potential dropped object events and provides a preliminary risk assessment for each accident. The detailed risk assessment step initiates with the dropped object events defined during the preliminary assessment step. Due to the large number of events likely to be identified, it is common to employ a screening approach to minimise the number of computationally expensive assessments. The initial screening removes low risk and extremely low likelihood events from further considerations. The remaining events are then examined using increasingly complex (and hence, less conservative) analysis techniques in addition to determining if mitigation alternatives are viable to reduce the risk or frequency.

The details for the key activities within the identification and assessment processes are presented in the following paragraphs.

4.2 Dropped Object Evaluation

4.2.1 Acceptance Criteria

The dropped object acceptance criteria for all affected systems should be defined by the asset owner. The acceptance criteria define the permissible amount of damage present in the system so that the risk levels for health and safety, environment, and facility can be assessed during and after the dropped object event.

The evaluation process considers the potential hazards associated with a dropped object impacting topsides structures and processing equipment and is not intended to address overboard/subsea impacts with fixed jacket and floating facility structure or subsea equipment from the risers/umbilicals to the wells themselves. In addition to the structural concerns, personnel health/safety as well as environmental aspects should be considered as appropriate. The acceptance criteria are to identify not only functional performance (e.g., acceptable structural damage to the element) but also time limits for safety critical elements so that crew is able to respond appropriately (e.g., muster and evacuate if required following the dropped object event).

The assessment of a dropped object focuses on the ability of a topsides structure (laydown areas, drill floor, material handling way, deck plating, and truss members) and associated processing equipment to absorb the dropped object's kinetic energy at the location of impact. For all but the smallest events, the primary structural effects associated with the dropped object are related to the plastic deformation of local members involved in the impact event and the potential for subsequent collapse following the impairment of the local members. The acceptance criteria should be based on the functional requirements of the element and the risk exposure of the event (incorporating both the frequency and consequence). The outcomes of the scenarios include health and safety risks, environmental releases, facility damage and down time.

It is envisioned that the acceptance of the damage will be based not only on the initial condition but the ability of the facility to function until the damage can be permanently repaired. Selection of the environmental conditions that the impaired facility must withstand will reflect the time to perform the repair. In lieu of additional information, ISO 19901-3 recommends the use of a 10-year environmental event.

All acceptance criteria and corresponding risk shall be defined by the asset owner and approved by Class prior to initiating the study.

4.2.2 Assessment Inputs

The dropped object assessment consists of four primary inputs:

(1) Dropped object scenario definition,
(2) Structural geometry,
(3) Material properties, and

(4) Applied loading.

These are described in the following paragraphs.

4.2.2.1 Dropped Object Scenario Definition

The dropped object scenarios considered are identified using the process outlined in Chap. 2. The scenarios considered may include events such as:

- Dropped objects during lifting operations from supply vessel,
- Swinging loads during crane operations, and
- Dropped objects during drilling operations.

The relative size, velocity and frequency of the dropped object events will be based on best available information for the site. It will include information such as facility general arrangement, anticipated (or actual) lifting activities, and anticipated (or actual) drilling operations. While historical information may be available, it is incumbent on the analyst to properly estimate the current and future (via trending) dropped object sizes and frequencies.

When selecting the scenarios to consider in more detail, it is assumed that a minimum annual likelihood considered will be utilised consistent with the structural design and accidents. The dropped object event definitions will be fully characterised by:

(1) On-going operations (e.g., lifting, drilling or severe environment)
(2) Dropped object description (similarly for floating facility):
   - Dropped object (including structure description and mass),
   - Drop characteristics (including location of anticipated impact and height of drop).
(3) Dropped object impact definition:
   - Velocity at impact,
   - Impact angle,
   - Location of impact,
   - Environmental conditions (e.g., weather and sea state),
   - Safety measures present (including structural and operational safeguards during the impact and safety/environmental barriers after event),
   - Likelihood of event.

4.2.2.2 Structural Configuration

The structural geometry considers the facility topsides structure and equipment. Sufficient information is required to develop structural models to assess the impact event. A local structural model will be developed at the impact point that is capable of capturing the high deformations, yielding, and failure of adjacent structural members. This local model will utilise a sufficiently large portion of the topsides structure so that boundary conditions become irrelevant or employ realistic boundary conditions from a global model. The extent of the local model will reflect the magnitude of the impacting event (i.e., larger extents will be utilised for higher energy events).

The definition will include particular attention to structures utilised as part of or influencing safety barriers (e.g., proximity to critical safety elements, personnel evacuation routes/muster areas, and degree to which structural damage can be tolerated).

4.2.2.3 Material Properties

The stiffness, strength, and ductility of individual structural components are the primary material properties associated with a dropped object assessment. Due to the dynamics involved, it may be appropriate to incorporate strain rate effects into the assessment. Selection of strain rate will vary based on dropped object scenario.

4.2.2.4 Applied Loading

The applied loads for a dropped object assessment address the various structural loading components present, namely:

- Dead loads,
- Live loads, and
- Dropped object loads.

Dead loads are composed of structural member and dry equipment weights. Live loads reflect the variable loads associated with the operation of the facility. The live loads selected should reflect the other design and accidental loading assessments performed. Definition of the impact load inputs will be defined via the initial hazard identification and assessment based on the dropped object mass, dimensions, and drop height. Environmental loads are not required when assessing the local effects of the dropped object as the structural response due to environmental loads during ordinary and lifting operations will be significantly less than those incurred from the dropped object itself. However, environmental loads may be required to evaluate global structural performance of the impaired facility including the hydrostatic (damage) stability of a floating structure in the event that the hull shell is ruptured or the global jacket strength of a damaged fixed platform (prior to remediation). In this case, specific post-event environmental loads should be defined.

4.2.3 Dropped Object Assessment

The dropped object assessment can be separated into two distinct phases. The first phase assesses the impact event on an individual component or region. The second phase considers the resulting damage influence on the global structural response.

The individual component assessment focuses on the direct structural damage caused by the dropped object. By ensuring conservation of energy, it is possible to equate the strain energy developed in the impacted structure (energy absorbed by both the dropped object and impacted structure/equipment) to the loss in kinetic energy of the dropped object. The typical assumption is that no energy absorption is present in the dropped object and that the topsides equipment and structure must have sufficient capacity to resist the incident kinetic energy. This is justified as the majority of dropped object studies focus on events such as axial impact from pipes where minimal energy absorption is observed in the dropped body.

There are two broad categories of methods to address the performance of the individual component: approximate static solutions (closed form equations and tabulated data) and nonlinear finite element analysis. Once the local damage has been quantified, the second phase is initiated, where its influence on the global facility performance is assessed, if required. This global assessment utilises traditional methods to predict the capacity of the topsides structure. It is noted that the damage of certain topsides equipment from a dropped object will typically only require a local damage assessment as such damage seldom initiates a global structural failure (exceptions exist, particularly during drilling operations when a loss of containment may result in a loss of well control).

Details pertaining to the assessment methods are presented in the following paragraphs.

4.2.3.1 Local Assessment: Static Approximate or Nonlinear FEA Methods

The local assessment relates the incident kinetic energy to the structural strain energy dissipation of a single structural member or unit. Using this approach, the mechanics can be expressed in terms of an energy balance relating the striking object kinetic energy to the strain energy stored in the deformed structures and any energy absorption barriers in place:

$$E_k = E_d + E_s + E_a \qquad (4.1)$$

where

E_k = kinetic energy associated with the dropped object at impact.

E_d
E_s = strain energy stored in and dissipated through elastic/plastic deformation of the dropped object and impacted structure, respectively.

E_a = energy absorption provided by safety barriers (if present).

As previously mentioned, the strain energy component of the dropped object may be neglect in certain circumstances, that is $E_d = 0$.

The kinetic energy present in the dropped object reflects is velocity at the instant of impact:

$$E_k = \frac{1}{2}(m)v^2 \qquad (4.2)$$

4.2 Dropped Object Evaluation

For incidents involving topsides equipment and structures, the velocity can be readily computed from the dropped object's height.

The next step in the local analysis is to develop the strain energy stored in the impacted structure such that the strain energy equals the incident kinetic energy minus the energy absorption capacity of the safety barriers. To do this, it is necessary to evaluate the energy dissipated via elastic and plastic deformations for the impacted structure. Depending on the energy level present during the event, these deformations may be complex and involve large plastic strains and potentially structural failures (e.g., plate/stiffener rupture or weld fracture). The two primary methods for predicting the resulting strain energy are approximate static solutions (closed form solutions or empirical tables/charts) and finite element analysis.

In some cases, it may be possible to assess the dropped object damage based on approximate static solutions in the form of existing closed form solutions or empirical tables or curves. Approximate static solutions provide a means to readily determine the deformations and associated strain energy for the impacted structure that equals the incident energy (kinetic minus absorption). The deformations are then used to assess the damage.

Nonlinear finite element analyses provide a means to accurately assess the impact mechanics while providing details on virtually every structural aspect of the event. Properly modelled nonlinear finite element simulations can capture many of the failure mechanisms anticipated for a dropped object event including material yielding, large deformations of plate and stiffener components, connection failures, and material rupture.

Note that it is unlikely that fracture of connections will be explicitly modeled during the assessment though it is possible to utilise representative strain or deformation limits to capture this failure mode. The modeling of the impact event typically requires a highly refined model in the vicinity of the contact area so as to capture not only the structural deformations but also the changing contact area between the bodies and the desired failure modes. The dynamic simulation will then trace the event from its initial contact through to the termination of the dropped object's velocity (assuming that all the incident velocity will be spent as strain energy and safety barrier energy absorption).

4.2.3.2 Global Assessment

Following development of the initial damage associated with the dropped object, it may be necessary to assess the impact on the global structural performance of the facility. This is typically only required when the potential for event escalation exists following the initial impact event (e.g., a primary load carrying member is crippled by the event). Assessment of these conditions will follow traditional assessment methods for determining the facility structural capacity and stability. However, the analyst must properly represent the damaged facility.

4.2.4 Mitigation Alternatives

Mitigation alternatives typically focus on modifying acceptable operational practices and reinforcing the facility's structure exposed to the potential impact events, including but not limited to:

- Modifying topsides equipment and structural member positions to reduce the likelihood of a dropped object during crane operations,
- Restricting drilling and crane operations where weather directly increases the frequency of dropped objects,
- Increasing structural protection (i.e., energy absorption) to equipment, and
- Some of these mitigations are only applicable to new designs (e.g., modifying the general arrangement).

4.2.5 Documentation

Documentation for the dropped object hazard evaluation process will address the following:

(1) Preliminary dropped object risk assessment (Fig. 4.1), and
(2) Detailed dropped object risk assessment (Fig. 4.2).

The preliminary dropped object risk assessment documentation provides a brief summary of the activities in Fig. 4.1:

- Acceptance criteria for structural and safety critical elements,
- Identified dropped object scenarios considered, and
- Preliminary risk assessment.

Where Class is requested to review the dropped object hazard evaluation process, it is prudent to provide the preliminary risk assessment to Class for review prior to initiating additional dropped object assessment activities. The assessment details the complete process capturing pertinent inputs and key structural outputs as presented in Fig. 4.2:

(1) Discussion of screening process for events requiring additional assessment,
(2) Description of the structural response as required by acceptance criteria,
(3) Identification of all required mitigation actions to for safe and efficient operations, and
(4) Detailed risk assessment.

5 Fire Hazards

5.1 General

The fire assessment provides an overview to the evaluation of potential fire events on offshore structures that could lead to personnel life/safety incidents, environmental release, and facility damage. The method presented is applicable to new designs as well as existing facilities, though the challenges faced with mitigating an event for an existing facility may be significantly more involved.

The fire event begins with accidental release of liquid or gaseous (or mixed) hydrocarbons into the environment followed by an ignition. Two types of fires are normally considered for offshore platforms: a pool fire and a jet fire. The pool fire develops when liquid inventory (e.g., liquid hydrocarbon) released on a deck forms a pool. The pool fire continues following ignition until either all the hydrocarbons are consumed or the ventilation conditions present cause the fire to be extinguished. In the case of a jet fire, a high-pressure release of gas or liquid hydrocarbons forms a jet that is subsequently ignited. The jet fire continues until the hydrocarbon source is consumed or the flame becomes unstable and extinguishes itself.

Thermal loads from either the pool or jet fire will increase the temperature of engulfed structural members as well as non-engulfed members near the fire via radiation, convection, and/or conduction. These structural members may experience degradation in both strength and stiffness as a function of exposure time due to the elevated temperature. The degradation of individual structures can then result in the entire structural system failing to meet its service requirements, ultimately manifesting into either a local or global failure.

The general evaluation process for fire hazards is as follows:

(1) Fire preliminary risk assessment (Fig. 5.1):
 - Define acceptance criteria,
 - Define fire events,
 - Perform fire assessment, and
 - Perform preliminary risk evaluation.
(2) Fire hazard detailed risk assessment (Fig. 5.2):
 - Select pertinent/applicable fire scenarios from the fire preliminary risk assessment,
 - Identify mitigation actions to meet assessment criteria or reassess the event using a more advanced (less conservative) analysis method,
 - Evaluate local and/or global structural and safety critical element performance, and
 - Evaluate risk exposure to determine if additional assessment or mitigations required.

The fire hazard identification process is similar to that presented in Chap. 2. However, there are some particulars regarding its execution that are addressed in this chapter, including commentary on the acceptance criteria definition (2.1), fire assessment inputs (2.2), and fire assessment (2.3).

5.1.1 Existing Standards for Reference

- API RP 2FB Recommended Practice for the Design of Offshore Facilities Against Fire and Blast Loading [2006],
- API RP 2A-WSD Recommended Practice for Planning, Designing, and Constructing Fixed Offshore Platforms—Working Stress Design: Section 18, Fire, Blast, and Accidental Loading [2008],
- ISO 13702 Petroleum and natural gas industries—Control and mitigation of fires and explosions on offshore production installations—Requirements and guidelines [1999],
- UKOOA/HSE Fire and Explosion Guidance [2007],
- Eurocode Eurocode 1: Actions on structures, Part 1–2: General actions—Actions on structures exposed to fire [2006],
- Eurocode 3: Design of steel structures, Part 1–2: General rules—Structural fire design [2005], and
- ISO 19901–3 Petroleum and natural gas industries—Specific requirements for offshore structures—Part 3: Topsides structure [2010].

5.1 General

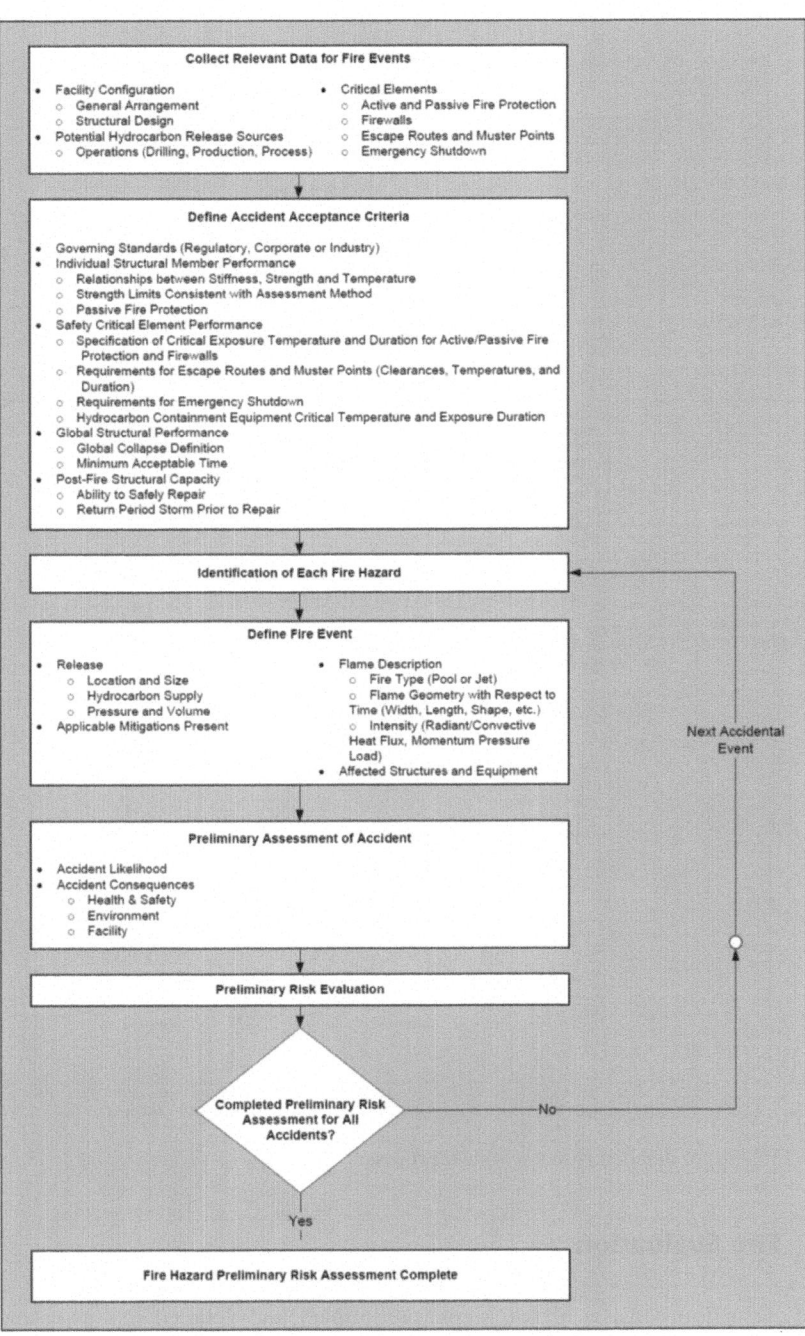

Fig. 5.1 Fire hazard preliminary risk assessment process

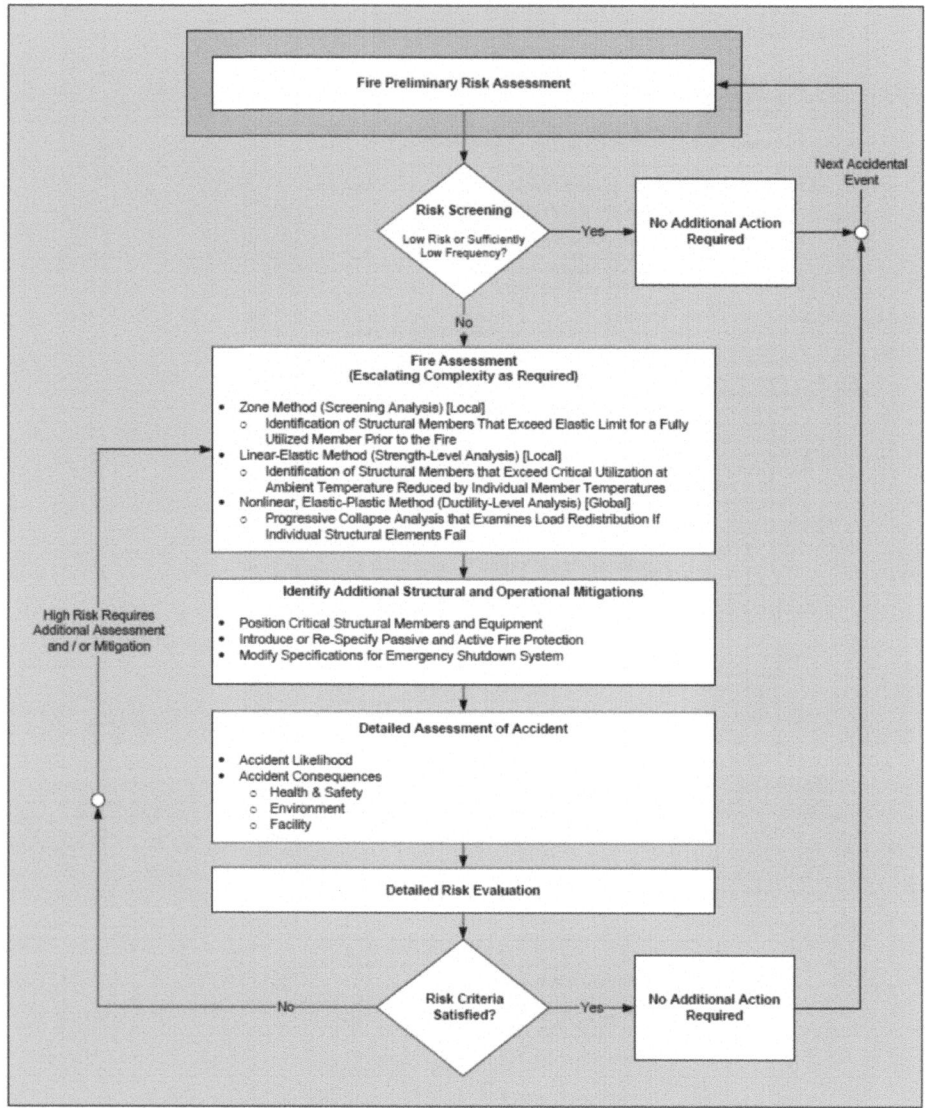

Fig. 5.2 Fire hazard detailed risk assessment process

5.2 Fire Evaluation

The fire evaluation consists of two steps: preliminary and detailed risk assessments, as presented in Chap. 4, Figs. 4.1 and 4.2, respectively. The preliminary risk assessment step, as presented in Chap. 2, defines the potential fire events, and provides a preliminary

5.2 Fire Evaluation

risk assessment for each accident. The detailed risk assessment step initiates with the fire events defined during the preliminary assessment step. Due to the large number of events likely to be identified, it is common to employ a screening approach to minimise the number of computationally expensive assessments. The initial screening removes minimal risk and extremely low likelihood events from considerations. The remaining events are then examined using increasingly complex (and hence, less conservative) analysis techniques in addition to determining if mitigation alternatives are viable to reduce the risk or frequency. The details for the key activities within the identification and assessment processes are presented in the following paragraphs.

5.2.1 Acceptance Criteria

Prior to performing any fire assessment, the acceptance criteria for all pertinent systems should be defined by the Owner. The acceptance criteria will define the permissible damage in the systems. This coupled with the damage assessment will characterise the risk levels for health and safety, environment, and facility during and after the fire event. The evaluation process considers not only the potential failure of individual elements (e.g., structural members or safety critical elements) but also the global structural performance.

The performance of the following shall be considered in the acceptance criteria development:

(1) Individual structural members,
(2) Safety critical elements, and
(3) Global structural system.

The acceptance criteria development for these three areas should include specification of a minimum functional time (e.g., one hour for personnel to muster and evacuate if appropriate) in addition to structural and thermal performance during the fire event. These may be governed by codes or regulatory requirements.

The primary effects of a fire on an individual structural member are the elevation of the member's temperature and thermal expansion. As the temperature increases, a reduction in its strength, stiffness and thermal expansion will be observed. The rate of material degradation depends on many factors including structural member material strength and stiffness at elevated temperatures, applied loading (thermal and structural), and the presence of active and passive fire protection. The member response to a fire may result in a variety of failures including (but not limited to) member yielding, buckling, formation of plastic hinges, exceeding deformation limits, connection failures, etc. Additional information about the development of individual structural failure criteria is detailed in Class, API, and ISO. It should be noted that, as detailed in Chap. 4, Sect. 2.3, the structural acceptance criteria will be dependent on the assessment method performed (i.e., performance measures reflect the relative conservatism present in the assessment method).

Safety critical elements encompass items intended to mitigate the consequences associated with a major fire event. These include firewalls, equipment necessary for the safe shutdown of the facility, personnel protection and escape, active fire suppression systems, hydrocarbon containment equipment, and critical communications equipment. Performance measures for all safety critical elements directly or indirectly affected by the fire scenario are to be considered so that personnel, environmental, and facility risks are managed consistently with the specified acceptable risk exposure. An example of this is the rating of firewalls exposed to pool fires. An A15 firewall is required to maintain its stability and integrity for 60 min with it taking 15 min for the temperature to reach 140 °C on the cold face of the wall. Any scenario that would exceed these limits would be identified as failing to pass the performance criteria.

The global structural performance incorporates the individual structural member and safety elements together to develop an overall performance measure focused on assessing structural collapse in a manner consistent with developed risk tolerance. Typically, two tiers of global performance are identified. The first considers local structural collapse following the fire event. A local collapse may be deemed acceptable provided it does not lead to an event escalation that introduces additional personnel or environmental consequences. The second considers a global structural collapse. In this event, the assessment evaluates the survivability of the facility. Consequences associated with a global collapse include not only those associated with the facility (above and below water level) but also the ability of personnel to evacuate the facility and potential environmental impact of the event.

Table 5.1 presents brief examples of fire acceptance criteria that may be utilised during the fire evaluation. In each case, explicit definition of the performance expectation is provided for all key structures and equipment that reflect health and safety, environmental, and financial risks.

All acceptance criteria and corresponding risk must be defined by the asset owner; if Class is requested to review the acceptance criteria and corresponding risk, these should be reviewed prior to initiating the study.

5.2.2 Fire Assessment Inputs

The fire assessment consists of four primary inputs:

(1) Fire scenario definition,
(2) Structural configuration,
(3) Material properties, and

Table 5.1 Example fire evaluation acceptance criteria

Structure/equipment		Acceptance criteria
Topsides structural members (criteria varies by screening method)	Zone method	All members must remain below 400 °C during the fire so that a maximum strain limit of 0.2% is maintained (highlights members that would exceed the elastic limit if they were fully utilised prior to the fire)
	Linear-elastic method	Acceptable structural utilisations are defined based on selected strain limit (e.g., 0.2%) and the reduction in yield stress with temperature (highlights members that exceed elastic limit based on utilisation prior to the fire and temperature during the fire)
	Nonlinear-elastic method	Local structural collapse (large plastic deformation and material/connection failure) is permitted provided that it does not lead to a global collapse or escalate the event in terms of personnel safety or environmental containment
Safety critical elements	Firewall	H60 firewall provides structural stability and integrity for at least 120 min with the cold face temperature taking 60 min to rise to 140 °C
	Personnel evacuation routes/muster areas	Evacuation routes are passable and muster areas are safe for 60 min after initiation of the fire event

(continued)

Table 5.1 (continued)

Structure/equipment		Acceptance criteria
	Processing pressure vessel	Impingement of the pressure vessel by jet fire does not induce loss of containment
	Accommodations	Within 60 min: • The temperature at any one point, including any joint, should not rise more than 180 °C above the original temperature, • The height of upper layer of smoke is not to exceed 1.8 m, • Visibility should not reduce to below 2 m (6.5 ft), and • Toxicity of CO should remain below 1,200 ppm When subjected to blast overpressure bulkheads, windows, doors, and similar penetrations are to remain intact. The accommodations are to retain sufficient structural integrity to prevent collapse

(4) Applied loading.

A brief overview of each input is presented in the following paragraphs.

5.2.2.1 Fire Scenario Definition

The fire scenarios are identified using the process outlined in Chap. 2. The scenarios identified for consideration capture pool and jet fire events with sufficient likelihood to represent a meaningful threat to the facility.

A target minimum likelihood should be determined during the hazard identification process so that only reasonable events are considered for further analysis. It should be noted that fire events may follow or precede an alternate accident. For instance, a fire event could be initiated by a dropped object, or a blast may follow a fire. The interactions between accidents should be addressed during the hazard identification process and, if required, within additional specialised assessments. While these joint events are outside the scope of this book, it is important that the appropriate path dependency for potential damage be followed so that potential damage is faithfully captured.

5.2 Fire Evaluation

The ultimate goal of the fire scenario definition is to be able to provide the thermal loading component for the assessment. As such, sufficient information to describe the event is required including:

(1) Release location,
(2) Release inventory (e.g., hydrocarbon supply),
(3) Assumed mitigations affecting fire event:
 - Isolation time (detection plus emergency shut down time),
 - Blow down time,
 - Active fire suppression systems.
(4) Fire description:
 - Release inventory (i.e., hydrocarbon supply),
 - Flame geometry with respect to time (width, length, shape, etc.),
 - Flame intensity:
   – Radiant heat flux,
   – Convective heat flux, and
   – Momentum pressure load.

Details pertaining to the fire description including fire geometry and flame intensity are captured in other documents such as API RP 2FB.

5.2.2.2 Structural Configuration

The structural geometry considers the facility's general arrangement and individual member configurations. The structural layout identifies the position of equipment and structural members relative to the release locations considered. This will influence the development of key factors pertinent to the acceptance criteria (e.g., proximity to critical safety elements, personnel evacuation routes/muster areas, and degree to which structural damage can be tolerated).

Individual structural member geometry definition is required to develop the individual structural member temperature profiles and the overall structural assessment. By utilising the fire event definition with the individual member geometry (shape and critical cross-sectional dimensions) and relative position to the fire, it is possible to develop the temperature profile for all affected members during the event.

5.2.2.3 Material Properties

The primary structural impact of a fire is on the strength and stiffness of the structural members. As the structure is assessed, it is critical that the analysis includes the degradation in strength and stiffness with respect to the fire. Various equations and tables are available for the performance of structural materials at elevated temperatures including strength, stiffness, and stress–strain behaviour. API RP 2FB and ISO 19901-3 provide guidance on the variation in key material properties of individual structural members with temperature.

Creep effects may be significant for steel members at elevated temperatures for a prolonged period of time. The temperature and duration of exposure to make creep significant can vary from 600 °C (ISO 19901-3) to as low as 300 °C for a highly stressed member. The impact of creep on the structural response should be considered especially for non-redundant, compression members at a critical temperature (e.g., 600 °C) for prolonged period of time. It is incumbent on the analyst to determine if the fire events warrant consideration of creep effects (i.e., definition of a critical exposure duration and temperature for structural members at which creep becomes significant).

5.2.2.4 Applied Loading

Loads are separated into two categories for the fire events:

(1) Thermal loads, and
(2) Structural loads.

Thermal loads are used to predict the individual member thermal histories given the fire scenario definition, the member's geometry, and the presence of any fire protection system (such as PFP) that may limit the heat flux into the structure. If the contribution of PFP is included, then any potential degradation in insulation properties with prolonged exposure to the fire is properly included. The thermal time histories for each exposed member are calculated based on:

- Fire intensity (incident radiant/convective heat fluxes, radiant/convective/conductive heat fluxes away from member),
- Proximity of member to flame (engulfed/non-engulfed member),
- Structure geometry (shape and dimensions), and
- Effect of fire protection system (if present) such as radiation attenuation by water spray system.

Details pertaining to computational methods to develop the thermal time histories are provided in documents such as API RP 2FB and ISO 13702.

Structural loads can be separated into several broad categories: dead loads, live loads, jet fire momentum loads, and environmental loads. Dead loads consist of structural member and dry equipment weights. Live loads reflect the variable loads associated with the operation of the facility. They may be taken either at 75% of their maximum design values or the value assumed during the facility fatigue assessment (API RP 2FB). In addition, functional loads such as helideck and crane loads may be considered but typically only if pertinent to the fire scenario identified. Jet fires may introduce a signification pressure load on structural members due to the momentum associated with the flame impinging on the structure's surface. Lastly, environmental loads may be included if pertinent to defining the structural stress state during and after the fire event. In each case, the applied loading should be consistent with the defined fire scenario.

5.2.3 Fire Assessment Methods

Fire event assessments are intended to integrate the individual structural and safety critical element performance into a global facility assessment. There are typically three different assessment methods utilised [API RP 2FB, ISO 19901-3, ISO 13702]:

(1) Zone method (screening analysis),
(2) Linear-elastic method (strength-level analysis), and
(3) Nonlinear, elastic–plastic method (ductility-level analysis).

The three methods represent an increasing level of sophistication. Typically, an analyst will progress through each method so as to eliminate a given fire scenario from further consideration due to structural, safety, and environmental performance being deemed acceptable per Sect. 2.1 with the least computational effort required. This results in a decreasing number of analyses being performed as the method complexity increases.

Regardless of structural analysis method selected, the facility needs to be verified to meet the acceptance criteria defined.

5.2.3.1 Zone Method (Screening Analysis)

The zone method is the simplest approach. The underlying premise to the zone method is that the structure is permitted to experience stresses up to yield during an event such as a fire. As the structural temperature increases, the yield strength will reduce. It is possible to relate the loss in strength to the elastic design utilisations so that acceptance is governed only by the maximum temperature for an individual member. For a 0.2% strain limitation, the yield strength for a steel member will be reduced to 0.60 its nominal value at 400 °C [API RP 2FB]. This reduced strength corresponds to the initial, non-accidental loading safety margin present in the design analyses.

Therefore, the individual structural member temperature is the only value required to perform a fire assessment according to the zone method.

The output from the zone method identifies members that would exceed the elastic limit for a fully utilised member prior to the fire event. The results from the zone method provide an Owner with two alternatives: apply additional mitigations so that no structural member exceeds its temperature threshold (e.g., 400 °C for a 0.2% strain limitation) or perform more refined analyses as presented in the following sections.

Several notes on the zone method:

(1) Higher strain limitations can be defined (thereby increasing the allowable temperature). However, the accompanying reduction in stiffness should be included for these analyses.
(2) Strength and stiffness reductions for different materials should be incorporated if applicable.

5.2.3.2 Linear-Elastic Method (Strength-Level Analysis)

A linear-elastic analysis is a strength level analysis to evaluate the structural performance considering both temperature profile and member utilisation. The scenarios may include either all potential fires highlighted for consideration by the hazard identification or only those fire events with elements that failed to pass the zone method. Depending on the maximum temperature profile attained by individual structural members for the duration of the fire, the reduced stiffness and yield strength of the member should be used in the structural analysis.

There are two primary inputs for the linear-elastic method:

(1) Structural member temperature profiles for the given fire scenario, and
(2) Structural utilisation levels prior to the fire.

Following selection of a strain limitation (e.g., 0.2%), the reduction in strength and stiffness can be established, as shown in Table 5.2. The assessment of a given fire scenario will determine the temperature of members affected by the fire and the degradation of the material properties. The structure is checked at the accidental limit state with appropriate load and resistance factors corresponding to the accidental limit state design. If the major framing of the structure does not pass the linear check, then one of three actions should be taken: (1) additional mitigations are required to reduce the member temperature, (2) reduce the member utilisation via either member load reduction or increasing member strength or (3) reassess the fire event using more advanced (less conservative) nonlinear elastic–plastic method.

The notes regarding the selection of higher strain limits and addressing strength and stiffness reductions for different materials mentioned for the zone method are also applicable for the linear elastic method.

5.2.3.3 Nonlinear, Elastic–Plastic Method (Ductility-Level Analysis)

The nonlinear, elastic–plastic method (ductility-level analysis) is the most refined analysis considered herein. It is a progressive collapse analysis method that allows for load redistribution as individual members fail due to the applied fire and structural loads. The load redistribution utilises alternative load paths to compensate for the lost strength in the failed or severely degraded member. In addition to the applied loads, the nonlinear analysis performed typically considers these inputs:

5.2 Fire Evaluation

Table 5.2 Relationship between maximum acceptable member utilisation at ambient temperature at maximum observed member temperature [API RP 2FB]

Maximum member temperature		Yield stress reduction factor	Maximum acceptable member utilisation at ambient
°C	°F		
400	752	0.60	1.00
450	842	0.53	0.88
500	932	0.47	0.78
550	1,022	0.37	0.62
600	1,112	0.27	0.45

- Structural member temperature time histories,
- Temperature dependent stress–strain curves, and
- Other temperature dependent material properties.

The ductility level analysis is usually performed in the time domain. The time history analysis traces the entire fire event from its initiation through to its termination. During the event, affected structural members experience both the applied structural static loads as well as the time varying thermal loads that degrade both strength and stiffness. The sequence of member failure and load redistribution should be captured in the analysis. The primary focus of the elastic–plastic analysis is to assess the structural system's maximum strength with minimum conservatism. As such, significant plastic deformations and damage associated with the fire event are often acceptable provided the damage does not cause the global failure and escalate the consequences of the event.

5.2.4 Mitigation Alternatives

During the course of the fire event assessments, it may be necessary to develop mitigations to meet the acceptance criteria. This may be achieved either by addressing the load or resistance portion of the assessment. The load portion is focused on identifying means to reduce the severity of the fire event itself, a process engineering solution. The resistance portion improves the structural response (structural engineering solution). The degrees to which these two alternatives can be applied to a new and existing facility may vary greatly.

Mitigation of the fire load severity can take a variety of forms. For instance, reducing the potential inventory for the fire may significantly reduce the duration, intensity, or extent of the event. It may also be possible to modify the placement of equipment to reduce or eliminate potential secondary release sources or isolate the release from critical elements (e.g., firewalls or deck plating). Design considerations for a firewall include not only the firewall rating but also its placement. A vast array of active fire suppression systems exist that may prove efficient at reducing the fire severity.

The primary means to improve the structural resistance to the fire event is with PFP to reduce the rate of heat flux into the structural member, thereby reducing the associated strength and stiffness degradations.

An alternative to the use of PFP may be to introduce firewalls or reduce the utilisation of the member either by providing alternate load paths or increasing its size.

5.2.5 Documentation

Documentation for the fire hazard evaluation process will address the following:

(1) Preliminary fire hazard risk assessment (Fig. 5.1), and
(2) Detailed fire hazard risk assessment (Fig. 5.2).

The preliminary fire hazard risk assessment documentation provides a brief summary of the activities presented in Fig. 5.1:

- Acceptance criteria for structural and safety critical elements,
- Identified fire scenarios considered,
- Preliminary risk assessment.

Where Class is requested to review the fire hazard evaluation process, it is prudent to provide the preliminary risk assessment to Class for review prior to initiating additional assessment activities. The assessment details the complete process capturing pertinent inputs and key structural outputs as presented in Fig. 5.2:

(1) Discussion of screening process for events requiring additional assessment,
(2) Description of the structural response as required by acceptance criteria,
(3) Identification of all required mitigation actions for safe and efficient operations, and
(4) Detailed risk assessment.

Blast Hazards

6.1 General

The blast assessment identifies and mitigates vulnerable components in offshore structures that could lead to personnel life/safety incidents, environmental releases and facility damage if exposed to potential blast events. The method presented is applicable to new designs as well as existing facilities, though the challenges faced with mitigating an event for an existing design may be significantly more involved.

The blast event begins with accidental release of gaseous hydrocarbons into the environment followed by an ignition source. The blast is characterised by a time-varying overpressure that consists of a peak pressure, rise time, and rebound duration. Loads on structural members and equipment should consider the effects of both overpressure due to the blast wave propagation and drag due to the flow of gas from the combustion process.

The evaluation process for blast hazards consists of the following actions:

(1) Blast preliminary risk assessment (Fig. 6.1):
- Define acceptance criteria,
- Define blast events,
- Perform blast assessment, and
- Perform preliminary risk evaluation.
(2) Blast detailed risk assessment (Fig. 6.2):
- Select pertinent/applicable blast scenarios from the blast preliminary risk assessment,
- Identify mitigation actions to meet assessment criteria or reassess the event using a more advanced (less conservative) analysis method,
- Evaluate local and/or global structural and safety critical element performance, and

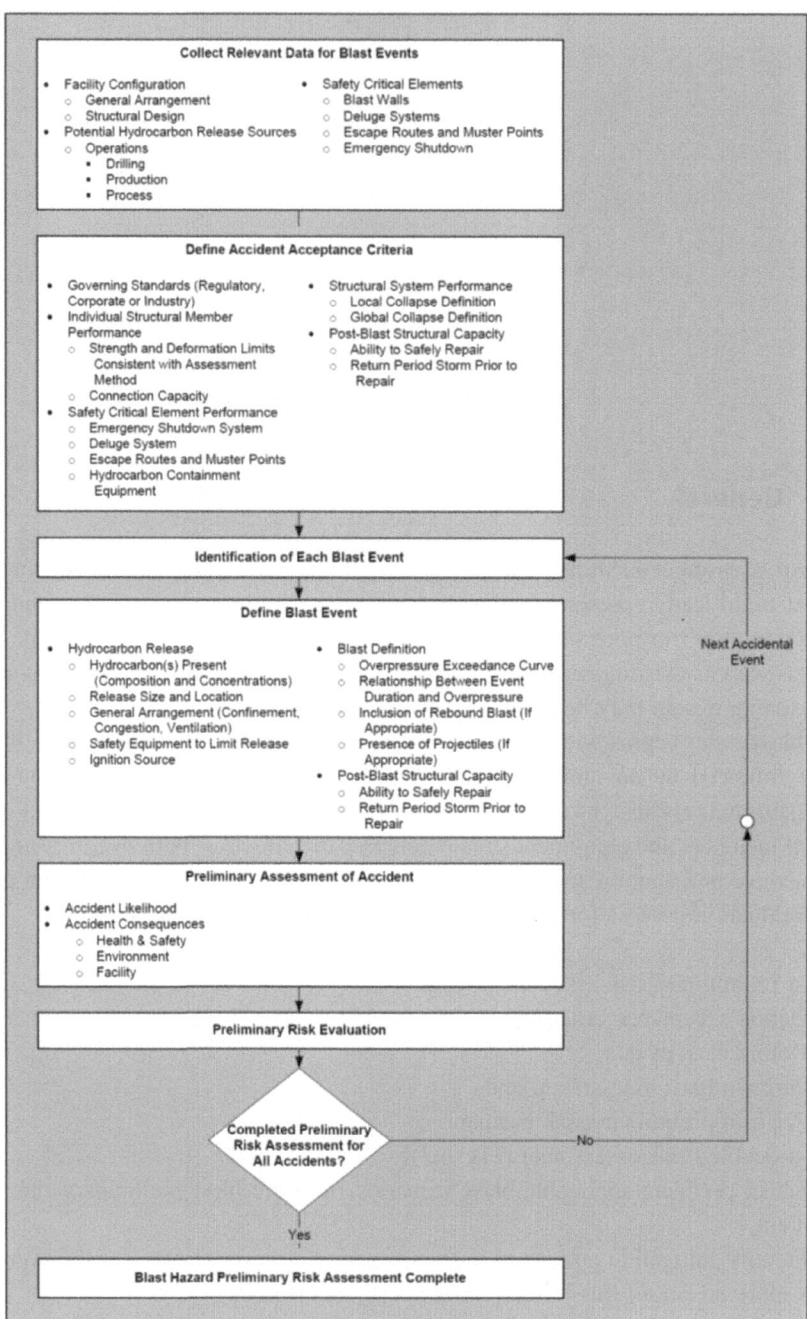

Fig. 6.1 Blast hazard preliminary risk assessment process

6.1 General

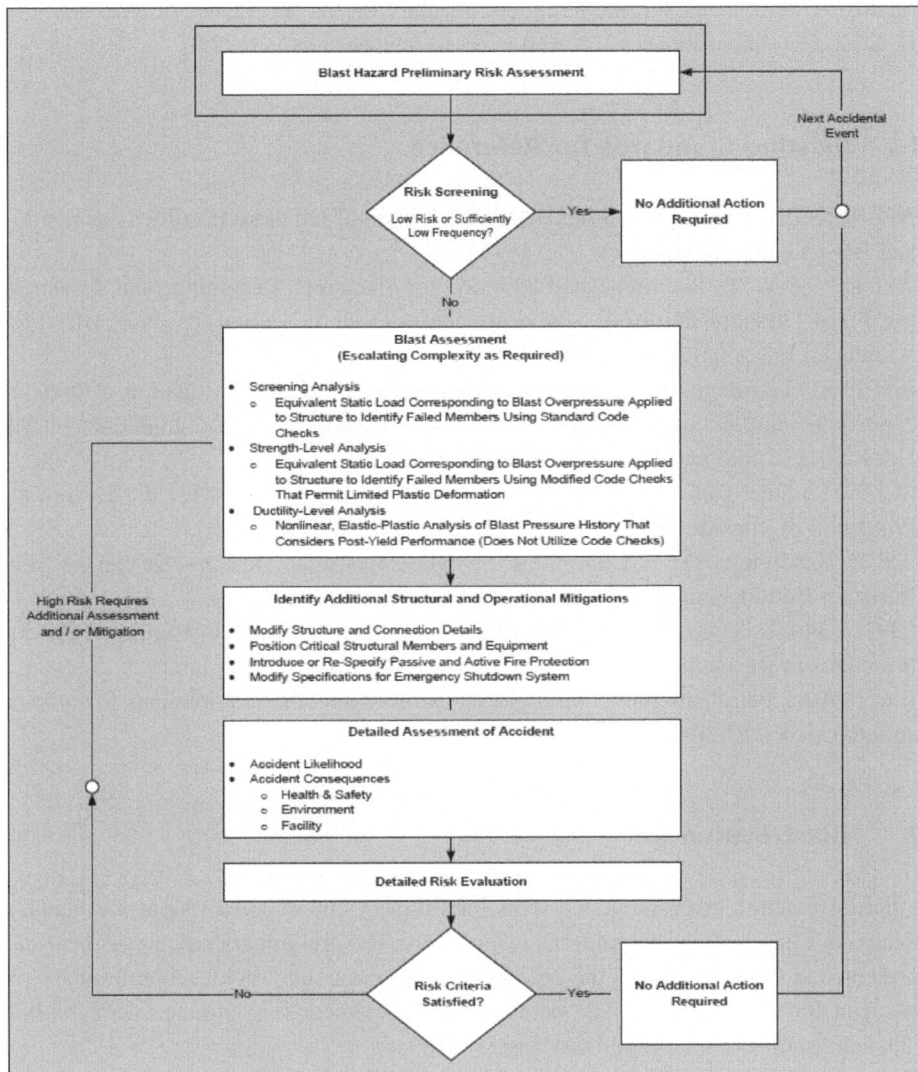

Fig. 6.2 Blast hazard detailed risk assessment process

- Evaluate risk exposure to determine if additional assessment or mitigations required.

While the blast hazard identification process is similar to that presented in Chap. 2, there are some specific areas regarding its execution that require additional clarification. These are addressed in this chapter, including commentary on the acceptance criteria definition (2.1), blast assessment inputs (2.2), and blast assessment methods (2.5).

6.1.1 Existing Standards for Reference

- API RP 2FB Recommended Practice for the Design of Offshore Facilities Against Fire and Blast Loading [2006],
- API RP 2A-WSD Recommended Practice for Planning, Designing, and Constructing Fixed Offshore Platforms—Working Stress Design: Section 18, Fire, Blast, and Accidental Loading [2008],
- ISO 13702 Petroleum and natural gas industries—Control and mitigation of fires and explosions on offshore production installations—Requirements and guidelines [1999],
- UKOOA/HSE Fire and Explosion Guidance [2007],
- SCI-P-112 Steel Construction Institute—Interim Guidance Notes for the Design and Protection of Topsides Structure against Explosion and Fire [1993],
- ASCE Handbook ASCE Committee on Blast Resistant Design—Design of Blast Resistant Buildings in Petrochemical Facilities [2010],
- UFC 3-340-02 United States Department of Defense "Unified Facilities Criteria: Structures to Resist the Effects of Accidental Explosions" [2008], and
- ISO 19901-3 Petroleum and natural gas industries—Specific requirements for offshore structures—Part 3: Topsides structure [2010].

6.2 Blast Evaluation

The blast evaluation consists of two steps: preliminary and detailed risk assessments, as presented in Chap. 4, Figs. 4.1 and 4.2, respectively. The preliminary risk assessment step, as presented in Chap. 2, defines the potential blast events, and provides a preliminary risk assessment for each accident. The detailed risk assessment step initiates with the blast events defined during the preliminary assessment step.

Due to the large number of events likely to be identified, it is common to employ a screening approach to minimise the number of computationally expensive assessments. The initial screening removes low risk and extremely low likelihood events from considerations. The remaining events are then examined using increasingly complex (and hence, less conservative) analysis techniques in addition to determining if mitigation alternatives are viable to reduce the risk or frequency. The details for the key activities within the identification and assessment processes are presented in the following paragraphs.

6.2 Blast Evaluation

6.2.1 Acceptance Criteria

The blast loading acceptance criteria for all pertinent systems should be defined by the asset owner prior to initiating the blast assessment. The acceptance criteria will define the permissible damage for the facility so that, when coupled with the blast assessment, it is possible to characterise the health and safety, environment, and facility risks during and after the blast event. The evaluation process considers not only the potential failure of individual elements (e.g., structural members) but also the global structural performance. The performance of the following shall be considered in the acceptance criteria development:

- Local structural members,
- Global structural system, and
- Critical systems that must remain functional directly following an event.

For individual members that provide support to flat surfaces (i.e., deck beams that support deck plate), the primary loads on individual members are due to the incident or reflected overpressure acting on the flat surface. Alternatively, demands on freestanding members will be dominated by drag loads. While the blast may introduce a thermal load component, the short duration of the event is typically insufficient to raise the member's temperature significantly unless a secondary fire is introduced by the explosion. The individual member may experience a wide variety of failure modes including overstressing, buckling, formation of plastic hinges, and connection failures, among others. Specifics concerning the development of individual structural failure criteria are detailed in ASCE Handbook and SCI-P-112. All failure modes for the structural component must meet the required response limit consistent with the analysis being utilised (refer to Chap. 5, Sect. 5.2.3). For example, connection details in a ductility-level analysis should be checked to verify that the anticipated deformations can be reached without premature failure (strength or deflection).

Safety critical elements encompass items intended to mitigate the consequences associated with a major blast event. These include both active and passive elements. The active safety critical systems are designed to address the sequence of events leading to the development of a hydrocarbon vapor cloud explosion. The passive safety systems focus on limiting inventory release and aid in personnel safety during the accident.

Specific examples include:

(1) Active safety systems:
   - Emergency Shutdown systems,
   - Blowdown system,
   - Deluge systems,
   - Gas detection system, and

- Alarm (requiring operator action).
(2) Passive safety systems:
  - Blast wall,
  - Fail-safe systems, and
  - Evacuation routes and muster areas.

An assessment of structural system performance incorporates both local and global structural components to develop an overall measure focused on structural damage and/or collapse. Typically, two tiers of structural performance are identified. The first tier considers local structural collapse caused by the blast event. Depending on the type and location of the structural component being considered, local collapse may be deemed acceptable provided it does not lead to an event escalation or introduce additional personnel or environmental consequences. The second tier considers a global structural collapse. In this event, the assessment evaluates the survivability of the facility. Consequences associated with a global collapse include not only those associated with the facility (above and below water level) but also the ability of personnel to evacuate the facility and potential environmental impact directly following the event.

Table 6.1 presents examples of acceptance criteria that may be utilised during the blast evaluation.

The topsides structural member acceptance criteria are intended to reflect the appropriate load and resistance factors utilised corresponding to the accidental limit state design. In each case, explicit definition of the performance expectation is provided for all key structures and equipment that reflect health and safety, environmental, and financial risks.

6.2.2 Blast Assessment Inputs

The blast assessment consists of four primary inputs:

(1) Blast scenario definition,
(2) Structural configuration,
(3) Material properties, and
(4) Applied loading.

A summary of each input is presented in the following paragraphs.

6.2.2.1 Blast Scenario Definition

The blast scenarios are identified during the hazard identification process outlined in Chap. 2. The blast assessment will typically involve two event sizes: (1) a higher likelihood but lower magnitude strength-level blast (SLB) and (2) a lower likelihood but higher magnitude ductility level blast (DLB). The appropriate blast levels should be selected by an asset owner so to be consistent with their overall risk tolerance for the facility. ISO

6.2 Blast Evaluation

Table 6.1 Example blast evaluation acceptance criteria

Structure/equipment		Acceptance criteria
Topsides structural members (criteria varies by assessment method)	Screening analysis	Equivalent static pressure is applied with the accidental limit state load and resistance factor checks. Actual material yield stress may be used in lieu of nominal yield stress
	Strength-level analysis	Equivalent static pressure is applied with the structure permitted to exceed utilisation ratio of 1.0 (actual material yield stress can be used) Specification of acceptance will consider the type of member loading (e.g., utilisation ratios of 2.5 for a tension member or 2.0 bending provided no buckling occurs)
	Ductility-level analysis	Nonlinear, elastic–plastic dynamic structural analysis. Standard code checks will not be applicable, so explicit definitions of acceptable limits are defined for: strength limits (0.8 ultimate stress), deformation limits (e.g., no impingement on safety critical elements, ductility ratio of 10 to limit damage, etc.), rupture or fracture defined by strain limits, and local buckling checks In addition to member checks, the connection details should be assessed to ensure that the structure is able to achieve the predicted member ductility
Safety critical elements	Blast walls	Blast wall will provide a barrier that shields a muster area from the blast overpressure

(continued)

Table 6.1 (continued)

Structure/equipment		Acceptance criteria
	Personnel evacuation routes/muster areas	Evacuation routes and muster areas will be clear of debris and provide safe/unobstructed passage
	Processing pressure vessel	Blast overpressure shall not pose an inventory containment issue for a hydrocarbon containment vessel and its associated piping

19901-3 identifies a minimum blast event likelihood of 10-4 (DLB) and a higher potential of 10-2 (SLB) as potentially requiring additional assessment.

The scenarios identified for consideration will capture a wide variety of factors required to assess the scenario including:

(1) Definition of hydrocarbon release:
- Release location, size, and direction,
- Type of hydrocarbon(s) present,
- Wind speed and direction,
- Dispersion of release,
- Release volume, composition, and concentrations, and
- All associated safety equipment present to limit release.

(2) Facility description pertinent to generation of blast description:
- Amount of structural confinement,
- Degree of equipment congestion,
- Presence of forced or natural ventilation,
- Active mitigation following the release prior to ignition, and
- Ignition source.

(3) Blast description:
- Overpressure exceedance curve,
- Relationship between event duration and overpressure (i.e., pressure curve),
- Inclusion of rebound blast (if appropriate), and
- Presence of projectiles (if appropriate).

(4) Facility description pertinent to assessing blast event consequences:
- Active safety systems,
- Passive safety systems,
- Occupied areas requiring protection, and
- Safety critical systems (i.e., BOP Control System, firewater pumps, etc.).

Precise description of the overpressure is outside the scope of this document. Development of blast overpressures for a confined/congested area can be achieved using three levels of models:

(1) Empirical Models: Overpressure correlated to experimental data (accuracy and applicability limited by model database),
(2) Phenomenological Models: Overpressure characterisation by incorporating physical principles into empirical observations (i.e., interpret observations so they are consistent with fundamental theory), and
(3) Numerical Models: Overpressure defined by solving the appropriate relationships for gas flow, combustion, and turbulence, typically utilising computational fluid dynamics principles.

Additional details pertaining to the blast event description are captured in other documents such as API RP 2FB.

It should be noted that blast events may follow or precede an alternate accident. For instance, a blast event could be initiated by a dropped object, or a fire may follow a blast. The interactions between accidents should be addressed during the hazard identification process and, if required, within additional specialised assessments. These joint events are outside the scope of this book.

6.2.2.2 Structural Configuration

The structural geometry considers both the general arrangement and individual member configurations. The structural layout identifies the position of equipment and structural members relative to the release locations considered. This will influence the development of key factors pertinent to the acceptance criteria (e.g., proximity to critical safety elements, personnel evacuation routes/muster areas, and degree to which structural damage can be tolerated). Individual structural member geometry definition is required for local and global damage assessments.

6.2.2.3 Material Properties

The primary concerns while assessing a blast event are the strength and stiffness of structural members and their connections. The short duration of the blast event means that there will be significant strain-rate effects on both the material strength and stiffness. However, there is insufficient time to develop a significant change in structural temperature, so thermal effects can be neglected. The utilisation of actual (as opposed to minimum specified) yield strength is typical for ultimate strength scenarios such as the ductility-level blast event. Dynamic strength increases for existing land-based blast assessments of buildings have been developed and should be applied so the facility may generate its full blast resistance capacity [ASCE Handbook, UFC 3-340-02].

6.2.2.4 Applied Loading

Loads are separated into two categories for the blast events:

(1) Blast loads, and
(2) Structural loads.

Blast loads will be based on information from the previously developed blast scenario definition for a given event. The blast loads will include the effects of both overpressure and drag. The blast overpressure will be defined by its distance from the ignition source, the blast exceedance curve (peak overpressure exceedance), and a pressure curve (generic example shown in Fig. 6.3 highlighting peak overpressure, rebound pressure, rise time, and blast event times). The drag loads are derived based on the pressure and gas flow velocity associated with the blast as well as the drag coefficients for the structure and equipment present. The global reaction loads result from the differential pressure loading of the facility.

Structural loads can be separated into several broad categories: dead loads, live loads, and environmental loads. Dead loads consist of structural member and dry equipment weights. Live loads reflect the variable loads associated with the operation of the facility. They may be taken either at 75% of their maximum design values or the value assumed during the facility fatigue assessment (API RP 2FB). In addition, functional loads, such as helideck and crane loads, may be considered but typically only if pertinent to the blast scenario identified. Lastly, environmental loads may be included if pertinent to defining

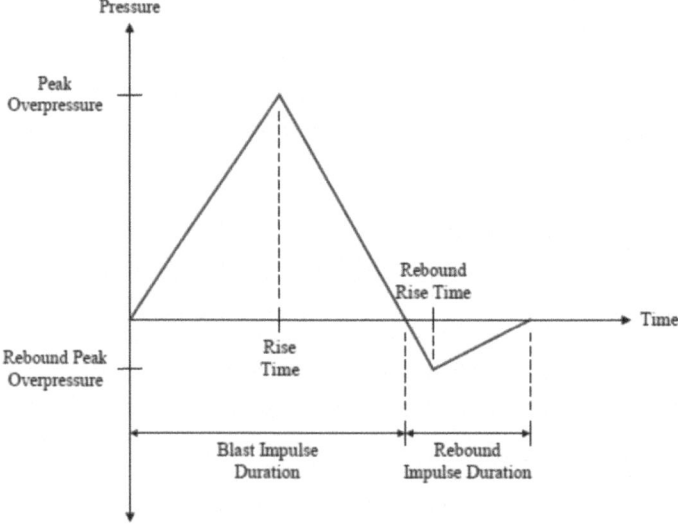

Fig. 6.3 Generic pressure curve highlighting key parameters

the structural stress state during and after the blast event. In each case, the applied loading should be consistent with the defined fire scenario.

6.2.3 Blast Assessment Methods

Blast event assessments are intended to integrate the individual structural and safety critical element performance into a global facility assessment. There are typically three different assessment methods utilised [API RP 2FB, ISO 19901-3, ISO 13702]:

(1) Screening analysis,
(2) Strength-level analysis, and
(3) Ductility-level analysis.

The three methods represent an increasing level of effort and accuracy (reduced conservatism). Typically, an analyst will progress through each method so to eliminate a given blast scenario from further consideration due to structural, safety, and environmental performance being deemed acceptable with the least computational effort required. This results in a decreasing number of analyses being performed as the method complexity increases.

While these three methods focus on the structural response, it is imperative that the analyst verify all safety critical elements function as intended for every blast scenario, regardless of analysis method selected.

In addition to assessing the defined blast scenarios identified during the hazard identification process, the analysts should investigate the robustness of the structural and safety systems. This is necessary due to the large amount of uncertainties present in the blast event definition and structural responses.

6.2.3.1 Single and Multiple Degrees of Freedom Model

Depending on the complexity of the structural system, the structure may be represented with either a single or multiple degree of freedom model. In general, the structural model should be representative of the area being analysed. Typically, individual member components can be represented using a single degree of freedom model and global response is captured using a multiple degree of freedom model.

Various types of single degree of freedom (SDOF) methods exist for analysing simple systems. They have the benefit of being quick to use but are limited because they are only applicable to very simple structural systems (e.g., a typical deck beam, a simple portal frame, etc.). Multiple degrees of freedom (MDOF) models are typically analysed using FEA software tools. The following guidelines should be followed when developing a MDOF model:

- Mesh size should be appropriate to capture global buckling effects (e.g., beam-column buckling),
- Section/material properties should be modified to properly account for the effects of local element buckling (e.g., local flange buckling in a deck beam). For some MDOF models, the mesh size may be fine enough to explicitly capture local buckling effects, but usually a global model is too large to allow for this level of refinement,
- Material properties should account for nonlinear material behaviour,
- Geometric nonlinearities should be captured (e.g., P-Delta effects),
- Strength increase factor should be used to account for the average or mean yield stress instead of the minimum specified design stress,
- Strain-rate effects on the material yield should be considered, and
- Dynamic effects should be fully captured.

6.2.3.2 Screening Analysis

The screening analysis is the simplest approach utilised to assess the structural response to a blast event. This method applies an equivalent static load and evaluates the response utilising accidental limit state design checks. The equivalent static load is the peak overpressure considered from the strength level blast scaled by a dynamic amplification factor. The structural members are permitted to experience member utilisation up to 1.0 with the actual material yield strength (including any yield stress increase due to strain-rate effects). Accidental limit state code checks are performed based on the calculated response to eliminate the event from consideration if no performance issues are identified for additional consideration.

6.2.3.3 Strength-Level Analysis

The strength-level analysis is a linear-elastic analysis of an equivalent static load corresponding to the blast overpressure that incorporates plastic code checks. As in the screening method, a dynamic load amplification factor is used to represent the overpressure peak. The code checks utilised during the analysis should reflect the basis of the blast overpressure. If the overpressure is derived from a blast exceedance curve directly, then utilisation in excess of 1.0 can be admitted as the structure is permitted to plastically deform (but not fail). The extent of the allowable utilisation will vary based on structural member type (compression or bending member, shear or bending support, etc.), but it is intended that the escalation will reflect the increased yield strength with strain rate as well as plastic deformation. If the SLB is defined to be 1/3 the DLB then no plastic deformation benefit should be claimed during the strength-level analysis (i.e., maximum utilisation is 1.0). In this case, the reserve capacity would be accounted for during a subsequent ductility analysis.

6.2.3.4 Ductility-Level Analysis

The nonlinear, elastic–plastic method (ductility-level analysis) is the most refined blast analysis method considered. It assesses the dynamic structural response when exposed to a blast event while accounting for geometric and material nonlinearities. It is anticipated that standard code checks will not be applicable to this assessment. As such, explicit checks should be performed to verify that the structural performance is within acceptable limits. This type of analysis assumes the structure will undergo plastic deformation and acceptance criteria would be set using deformation limits, strain limits (i.e., steel rupture), potential for buckling, connection capacities, etc. Ultimately, the goal of the analysis is to ensure that the structure will not catastrophically collapse under the blast load and that no local collapse escalates the personnel health and safety or environmental risk exposure.

6.2.4 Mitigation Alternatives

Mitigation alternatives for blast events are varied and cover structural design, process design, and the use of active and passive protection systems. The structural design mitigations include the placement of equipment/structures (including blast walls), ventilation, utilisation of blast walls and relief panels, and utilisation of proper design detailing so as to maximise ability of the structure to develop its full plastic capacity. The process design focuses on minimising the likelihood of hydrocarbon release that could fuel the explosion and subsequent ignition threats. The active and passive safety measures include various sensors such as gas detection prior to the blast, deluge systems upon gas detection, and the ability to quickly shutdown the facility upon release so as to minimise the likelihood of escalation.

6.2.5 Documentation

Documentation for the blast hazard evaluation process will address the primary activities associated with:

(1) Preliminary blast risk assessment (Fig. 6.1), and
(2) Detailed blast risk assessment (Fig. 6.2).

The preliminary blast risk assessment documentation provides a summary of the activities presented in Fig. 6.1:

- Acceptance criteria for structural and safety critical elements,
- Identified blast scenarios considered, and
- Preliminary risk assessment.

Where Class is requested to review the blast hazard evaluation process, it is prudent to provide the preliminary risk assessment to Class for review prior to initiating additional assessment activities. The assessment details the complete assessment process capturing pertinent inputs and key structural outputs as presented in Fig. 6.2:

(1) Discussion of screening process for events requiring additional assessment,
(2) Description of the structural response as required by acceptance criteria,
(3) Identification of all required mitigation actions for safe and efficient operations, and
(4) Detailed risk assessment.

Correction to: Accidental Load Analysis and Ensign for Offshore Structures

Correction to:
A. A. Olsen, *Accidental Load Analysis and Design for Offshore Structures,* **Synthesis Lectures on Ocean Systems Engineering, https://doi.org/10.1007/978-3-031-74773-1**

This book contains overlap in text with the previously published content [1] that was inadvertently omitted. The authors failed to attribute the reference [1]. The authors have now obtained permission to re-use this content from the American Bureau of Shipping.

Where [1] is: American Bureau of Shipping (2024), Rules and Guides https://ww2.eagle.org/en/rules-and-resources/rules-and-guides.html

---

The updated version of this book can be found at
https://doi.org/10.1007/978-3-031-74773-1

Glossary

DLB *Ductility-Level Blast.* DLB is a lower frequency, higher severity blast event.
ESD *Emergency Shutdown.* ESD is a system to shut in production systems in the event of a threatening situation.
FMEA *Failure Modes and Effects Analysis.* FMEA examine equipment or structure to identify single mode failures (i.e., what causes the equipment or structure to fail) and define the subsequent effect of the failure on the facility.
FPSO *Floating Production, Storage, and Offloading.* FPSO is floating oil and gas facility with a ship hull form.
HAZOP *Hazard Operability.* HAZOP analysis is a formal method to identify critical areas associated with the operability of the facility.
HSE *Health and Safety Executive.* The HSE is a public body in the United Kingdom responsible for the regulation and enforcement of workplace health, safety, and welfare, and for research into occupational risks.
MDOF *Multiple Degrees of Freedom.* MDOF is a simplified model capable of capturing the blast response of built-up structural members.
PFP *Passive Fire Protection.* PFP aims to contain fires or slow the spread of fires, through the design of fire-resistant structures. This includes the use of fire barriers (firewalls, fire doors, windows, etc.) and structural fire protection such as fireproofing structural material or coating.
SDOF *Single Degree of Freedom.* SDOF is a simplified model capable of capturing the blast response of single structural members.
SLB *Strength-Level Blast.* SLB is a higher frequency, lower severity blast event.

References and Publications

API RP 14J Recommended Practice for Design and Hazards Analysis for Offshore Production Facilities.
API RP 2A-WSD Recommended Practice for Planning, Designing, and Constructing Fixed Offshore Platforms – Working Stress Design: Section 18, Fire, Blast, and Accidental Loading [2008].
API RP 2FB Recommended Practice for the Design of Offshore Facilities Against Fire and Blast Loading [2006].
ASCE Handbook ASCE Committee on Blast Resistant Design – Design of Blast Resistant Buildings in Petrochemical Facilities [2010].
ISO 13702 Petroleum and natural gas industries – Control and mitigation of fires and explosions on offshore production installations – Requirements and guidelines [1999].
ISO 19901-3 Petroleum and natural gas industries – Specific requirements for offshore structures – Part 3: Topsides structure [2010].
SCI-P-112 Steel Construction Institute - Interim Guidance Notes for the Design and Protection of Topsides Structure against Explosion and Fire [1993].
UFC 3-340-02 United States Department of Defense "Unified Facilities Criteria: Structures to Resist the Effects of Accidental Explosions" [2008].
UKOOA/HSE Fire and Explosion Guidance [2007].

The manufacturer's authorised representative in the EU is Springer Nature Customer Service Centre GmbH, Europaplatz 3, 69115 Heidelberg, Germany. If you have any concerns regarding our products, please contact ProductSafety@springernature.com

Printed and bound by CPI Group (UK) Ltd, Croydon, CR0 4YY

26/03/2026

02078952-0020